T0141887

Analog Circuits and Signal Processing

Series Editors:

Mohammed Ismail, Dublin, USA

Mohamad Sawan, Montreal, Canada

The Analog Circuits and Signal Processing book series, formerly known as the Kluwer International Series in Engineering and Computer Science, is a high level academic and professional series publishing research on the design and applications of analog integrated circuits and signal processing circuits and systems. Typically per year we publish between 5–15 research monographs, professional books, handbooks, edited volumes and textbooks with worldwide distribution to engineers, researchers, educators, and libraries.

The book series promotes and expedites the dissemination of new research results and tutorial views in the analog field. There is an exciting and large volume of research activity in the field worldwide. Researchers are striving to bridge the gap between classical analog work and recent advances in very large scale integration (VLSI) technologies with improved analog capabilities. Analog VLSI has been recognized as a major technology for future information processing. Analog work is showing signs of dramatic changes with emphasis on interdisciplinary research efforts combining device/circuit/technology issues. Consequently, new design concepts, strategies and design tools are being unveiled.

Topics of interest include:

Analog Interface Circuits and Systems;

Data converters;

Active-RC, switched-capacitor and continuous-time integrated filters;

Mixed analog/digital VLSI;

Simulation and modeling, mixed-mode simulation;

Analog nonlinear and computational circuits and signal processing;

Analog Artificial Neural Networks/Artificial Intelligence;

Current-mode Signal Processing;

Computer-Aided Design (CAD) tools;

Analog Design in emerging technologies (Scalable CMOS, BiCMOS, GaAs, heterojunction and floating gate technologies, etc.);

Analog Design for Test;

Integrated sensors and actuators;

Analog Design Automation/Knowledge-based Systems;

Analog VLSI cell libraries;

Analog product development;

RF Front ends, Wireless communications and Microwave Circuits;

Analog behavioral modeling, Analog HDL.

More information about this series at http://www.springer.com/series/7381

Muhammad Yasin • Jeyavijayan (JV) Rajendran
Ozgur Sinanoglu

Trustworthy Hardware Design: Combinational Logic Locking Techniques

 Springer

Muhammad Yasin
New York University Abu Dhabi
Saadiyaat Island, Abu Dhabi
United Arab Emirates

Jeyavijayan (JV) Rajendran
Department of Electrical & Computer
Engineering, WEB 333H
Texas A&M University
College Station, TX, USA

Ozgur Sinanoglu
New York University Abu Dhabi
Saadiyaat Island, Abu Dhabi
United Arab Emirates

ISSN 1872-082X   ISSN 2197-1854   (electronic)
Analog Circuits and Signal Processing
ISBN 978-3-030-15336-6   ISBN 978-3-030-15334-2   (eBook)
https://doi.org/10.1007/978-3-030-15334-2

This Springer imprint is published by the registered company Springer Nature Switzerland AG.
The registered company address is: Gewerbestrasse 11, 6330 Cham, Switzerland

Muhammad Yasin would like to dedicate this book to the memory of his loving father and Prof. Masood Ahmad.
Jeyavijayan (JV) Rajendran would like to dedicate this book to his foreparents and teachers.
Ozgur Sinanoglu would like to dedicate this book to Cansu, Batu, and Bora.

Foreword

Today, we live in a world of constant communication, streaming entertainment, self-driving cars, and artificial intelligence all made possible by microelectronics and integrated circuits (IC). National defense enjoys similar benefits from seamless distributed military operations to smart weapons and electronic warfare. Microelectronics and integrated circuits are central to both our way of life and our defense of it.

As the role of microelectronics has grown, the way we manufacture them has changed. In the past, the Department of Defense accounted for roughly 40% of the microelectronics market; now, it is less than 0.5%. Due to the fantastic competition and cost pressures, what was a mostly on-shore industry dominated by American companies has become a global enterprise with a world wide supply chain. In fact, many of the world's largest electronic companies, such as Apple, are fabless. To provide the best electronics at the lowest prices, they outsource all their manufacturing to hyper-specialized on- and off-shore companies. This has been an unmitigated boon to consumers but presents risks both to fabless companies and nations. These risks include, but are not limited to, hardware-level tampering, reverse-engineering and intellectual property piracy, overproduction, counterfeiting, etc. For nation-states, the implications of these risks are even more dire especially for integrated circuits deployed in security-critical applications.

The key to an adversary realizing the aforementioned risks is a successful reverse engineering of the electronic part. Therefore, protecting the hardware design is central to any defensive strategy. A promising solution to protect the hardware is logic locking. Locking a circuit from reverse engineering with cryptographic-like rigor seems to be the most effective way to protect hardware against current and future attacks.

In 2017, I started the DARPA's Obfuscated Manufacturing for GPS program to seek out, develop, and validate the best logic locking techniques in academia and industry. The authors of this book formed the core of one of the most successful teams on the program. Their important 2012 work defined the threat model that has been used since then for logic locking and developed the first attack on logic locking.

This book provides a chronological evolution of the logic locking field. It starts with the very early logic locking defenses and attacks upon them and follows their descendants on both attack and defense to the present day. In the process, the book defines security metrics, different classes of attacks, and describes various design methods to deliver resiliency against these attacks. It serves not only as a detailed logic locking survey but also as a primer, with simple-to-follow examples for practicing designers.

The authors of this book have first-hand experience in the painstaking iteration from defense to offense and back again that are the hallmark of a robust logic locking technique. This familiarity with both sides of the attack/defense makes them the right team to deliver this highly informative book on logic locking. I am confident that it will be of great value to both academia and industry.

DARPA Microsystems Technology Office   Ken Plaks
Washington, DC, USA
May 2019

Preface

The evolving complexity of integrated circuits and the skyrocketing cost of owning and maintaining foundries have spawned the growth of fabless business model. Fabless companies concentrate their resources and efforts on product design and marketing and outsource the complex fabrication, test, and assembly processes to offshore foundries and test/assembly companies that specialize in these services. This outsourcing of services to global vendors makes it easier for untrusted entities to gain access to proprietary assets and even manipulate the processing steps. Consequently, the globalization of the IC supply chain has led to the emergence of multiple security threats such as IP piracy, overbuilding, counterfeiting, reverse engineering, and the insertion of hardware Trojans. Apart from the direct economic losses to the industry, these security vulnerabilities pose a severe risk to the safety and reliability of electronic systems, not barring even life-threatening scenarios.

Many countermeasures have been developed and deployed to mitigate the security threats and rebuild the trust in a globalized supply chain. These defenses include watermarking, split manufacturing, camouflaging, hardware metering, and logic locking. A common challenge for these countermeasures is to offer the maximum security at the minimal implementation cost and with minimal changes to the conventional IC design/fabrication process. Logic locking has emerged as the most promising, versatile, and easy-to-integrate solution among all the aforementioned defenses. By incorporating simple additional logic into a circuit during the design phase, logic locking can mitigate security against IP piracy, overbuilding, and reverse engineering.

Over the last decade, logic locking has been garnering increasing interest from the research community, including academia and industry. The continued emergence of different classes of logic locking defenses and attacks has led to ever-stronger defenses, raising the bar for the attackers. The authors of this book have been involved with the logic locking research since its inception, with the involvement spanning from the publication of multiple fundamental papers in this field to the development of the first logic locked chip. The significant advancements in this new field and the increasing interest of the research community have motivated multiple systematization-of-knowledge attempts in the form of book chapters and journal papers of tutorial nature. However, such publications are typically a recompilation of one or more research papers, with a focus on summarizing the state-of-the-art research. To attract practitioners to the field of logic locking, and thus hardware security, there has been a dire need to convey the fundamental principles following a pedagogical approach.

This book is an attempt to cover both breadth and depth of logic locking. It presents a comprehensive summary of logic locking defenses and attacks, describes their fundamental principles, and highlights the important research results. Consistent with the research trends, the bulk of the book is dedicated to the countermeasures that defend against the powerful SAT attack. The book systematizes the knowledge on logic locking attacks and defenses. It groups similar attacks and defenses, explains the common principles in detail, and elaborates on the essential differences. It supplements every important concept with illustrative circuit examples.

The book contains 11 chapters and an appendix on VLSI testing. Each chapter has been planned to emphasize the fundamental principles behind different classes of logic locking attacks and algorithms, progressively relating the new concepts to the previous ones. The first two chapters are introductory in nature, defining logic locking and presenting its brief history followed by a classification of attacks and defenses. We classify logic locking defenses into two main classes, pre-SAT and post-SAT, and the attacks into four classes, algorithmic, approximate, removal, and side-channel. Each of the following chapters focuses on a specific attack/defense technique. The last chapter summarizes the approaches presented in the book, highlights their challenges, and presents a few future research directions. Below is a comprehensive description of the contents of each chapter:

Chapter 1 motivates the need of logic locking in the context of the existing security vulnerabilities, introduces the basic definitions associated with logic locking, and compares logic locking with other design-for-trust approaches.

Chapter 2 presents a comprehensive history of logic locking followed by a classification of the logic locking attacks and defenses. It also introduces the metrics used to evaluate various logic locking approaches.

Chapter 3 focuses on the earliest "pre-SAT" logic locking techniques and presents three locking techniques, random, fault analysis-based, and strong logic locking, which essentially select suitable locations for inserting the additional logic, i.e., the XOR/XNOR or MUX key gates, into a netlist. It also introduces the sensitization attack that leverages the principles of VLSI testing to extract individual key bits of the secret key.

Chapter 4 focuses on the SAT attack, which is the most powerful of all attacks mounted on logic locking and circumvents all pre-SAT logic locking techniques. The attack is based on the notion of Boolean satisfiability. The chapter includes a review of the fundamental concepts of Boolean satisfiability, in addition to describing the attack algorithm with illustrative examples.

We classify the post-SAT logic locking techniques, which aim mainly at thwarting the SAT attack, into three subclasses: (1) point function-based, (2) SAT-unresolvable structure-based, and (3) stripped functionality-based logic locking

(SFLL). Chapter 5 focuses on the first subclass, which resists the SAT attack by increasing the number of SAT attack iterations. The chapter introduces three-point function-based logic locking techniques: (1) SARLock that integrates a comparator with the original design, (2) Anti-SAT that utilizes two complementary Boolean functions, and (3) AND-tree detection that searches for and locks the point functions already present in the original netlist.

While the point function-based locking techniques exhibit significant resilience to the SAT attack, these techniques and their variants remain vulnerable to two types of attacks: the approximate attacks and the removal attacks. The next two chapters of the book focus on these two classes of attacks. Chapter 6 elaborates on the operation of the approximate attacks that target compound locking techniques that integrate pre- and post-SAT locking techniques. It also presents two attacks, namely, AppSAT and Double-DIP, which recover only an approximate key.

Chapter 7 describes removal attacks that rely on the structural properties of various implementations of point functions to identify and isolate the original netlist from the protection circuitry. It presents four removal attacks, with each attack targeting a different point function-based defense.

The next two chapters of the book discuss the remaining two subclasses of post-SAT logic locking techniques. Chapter 8 describes how the SAT attack can be thwarted by inserting special structures that are hard to resolve by a SAT solver. The chapter introduces cyclic logic locking and one-way function-based logic locking (ORF-Lock). While cyclic logic locking can be broken using the CycSAT attack, ORF-Lock incurs high implementation cost and also remains vulnerable to removal attacks.

Chapter 9 presents SFLL, where the idea is to implement a modified on-chip circuit and restore the original functionality of the chip only upon activation with the correct key. SFLL is the first technique to offer quantifiable protection against all classes of attacks.

Chapter 10 elaborates on the side-channel attacks that leverage physical channels such as power and timing to extract information about the secret key. The chapter presents four side-channel attacks: the differential power analysis attack, the test-data mining attack, the hill-climbing attack, and the de-synthesis attack.

We anticipate the primary audience of the book to be the senior/graduate students in electrical and computer engineering and professionals in IC design and CAD software development, who have at least a rudimentary familiarity with the IC design flow. This book can be used as a textbook for courses on hardware security, VLSI CAD, or IC design. It can also serve as a "designer's guide" to implement logic locking in hardware designs. The book introduces the basic concepts of logic locking systematically in a way easy to follow for readers new to this field.

College Station, TX, USA   Muhammad Yasin
College Station, TX, USA   Jeyavijayan (JV) Rajendran
Abu Dhabi, United Arab Emirates   Ozgur Sinanoglu
Jan 2019

Acknowledgments

The authors would like to thank Prof. Ramesh Karri for his support and encouragement to pursue this effort. They would also like to acknowledge Zhaokun Han for proofreading and verifying the examples used in this book. The research covered in this book has in part been supported by the National Science Foundation (NSF), Semiconductor Research Corporation (SRC), Army Research Office (ARO), Defense Advanced Research Projects Agency (DARPA), Mubadala-SRC Center of Excellence for Energy-Efficient Systems (ACE4S), NYU/NYUAD Center for Cyber Security (CCS), and Texas A&M University. The authors acknowledge the crucial role of all these agencies and institutes in enabling this effort.

Contents

1   **The Need for Logic Locking** 1
    1.1   Globalization of the IC Design 2
        1.1.1   Traditional IC Design Flow 2
        1.1.2   Globalized IC Design Flow 2
    1.2   The Emergence of Hardware Security Vulnerabilities 4
        1.2.1   IP Piracy 4
        1.2.2   Overbuilding 4
        1.2.3   Hardware Trojans 4
        1.2.4   Counterfeiting 5
        1.2.5   Reverse Engineering 6
    1.3   Design-for-Trust (DfTr) Solutions 6
        1.3.1   Watermarking and Fingerprinting 7
        1.3.2   Camouflaging 7
        1.3.3   Metering 8
        1.3.4   Logic Locking: An Overview 8
        1.3.5   Logic Locking vs. Other DfTr Techniques 9
    1.4   Logic Locking: Definitions and Terminology 10
        1.4.1   IC Design Flow with Logic Locking 10
        1.4.2   Terminology 12
        1.4.3   Protection Against Hardware-Based Attacks 13
    1.5   Takeaway Points 13
    References 14

2   **A Brief History of Logic Locking** 17
    2.1   Milestones in Logic Locking 17
        2.1.1   The First Defense 18
        2.1.2   The First Threat Model and Attack 18
        2.1.3   The Most Powerful Attack 19
    2.2   Classification of Attacks and Defenses 19
        2.2.1   Classifying Logic Locking Techniques 19
        2.2.2   Classifying Attacks on Logic Locking 20
        2.2.3   A Timeline of Logic Locking 20

2.3   An Overview of Existing Defenses   21
    2.3.1   Pre-SAT Logic Locking   21
    2.3.2   Post-SAT Logic Locking   23
2.4   Logic Locking Attacks   25
    2.4.1   Algorithmic Attacks   25
    2.4.2   Structural Attacks   25
    2.4.3   Side-Channel Attacks   26
2.5   Attack-Defense Matrix   26
2.6   The Ever-Evolving Metrics   28
References   29

3   Pre-SAT Logic Locking   33
3.1   Random Logic Locking   33
    3.1.1   Motivation   33
    3.1.2   The RLL Algorithm   34
3.2   Fault-Analysis Based Logic Locking   36
    3.2.1   Motivation: Black-Box Usage   36
    3.2.2   Logic Locking and Fault Analysis   36
    3.2.3   The FLL Algorithm   37
3.3   Sensitization Attack   39
    3.3.1   Threat Model   39
    3.3.2   Attack Algorithm   40
3.4   Strong Logic Locking   42
    3.4.1   Basic Idea   42
    3.4.2   Pairwise Security   42
    3.4.3   The SLL Algorithm   45
3.5   The Variants of Basic Techniques   45
References   46

4   The SAT Attack   47
4.1   Preliminaries   47
    4.1.1   Boolean Satisfiability   47
    4.1.2   Tseitin Transformation   48
    4.1.3   Miter Circuit   49
4.2   The SAT Attack   50
    4.2.1   Distinguishing Input Patterns (DIPs)   50
    4.2.2   Attack Algorithm   51
4.3   Effectiveness Against Pre-SAT Logic Locking   53
4.4   How to Thwart the SAT Attack?   54
4.5   Formal Security Analysis Framework   54
References   55

5   Post-SAT 1: Point Function-Based Logic Locking   57
5.1   Maximizing SAT Attack Resilience   57
    5.1.1   Strong and Weak DIPs   57
    5.1.2   Circuits that Generate Weak DIPs   58

	5.2	SARLock	59
		5.2.1 Architecture	59
		5.2.2 Security Analysis	60
	5.3	Anti-SAT..	61
		5.3.1 Architecture	61
		5.3.2 Security Analysis	62
		5.3.3 Functional and Structural Obfuscation	63
	5.4	AND-Tree Detection	64
		5.4.1 Security Analysis	64
	5.5	A Comparative Analysis	66
	5.6	The Common Pitfalls....................................	66
	References		67
6	**Approximate Attacks**...		69
	6.1	Introduction...	69
		6.1.1 Compound Logic Locking	69
		6.1.2 Approximate Attacks	70
	6.2	AppSAT...	71
		6.2.1 Basic Idea	71
		6.2.2 Termination Criterion	72
		6.2.3 Random Query Enforcement............................	72
		6.2.4 Attack Algorithm	73
	6.3	Double-DIP...	73
		6.3.1 Basic Idea	73
		6.3.2 2-DIPs	74
		6.3.3 Attack Algorithm	75
	References		76
7	**Structural Attacks**		77
	7.1	Signal Probability Skew (SPS) Attack	77
		7.1.1 Basic Idea	77
		7.1.2 Preliminaries: Signal Probability Skew.....................	78
		7.1.3 Attack Algorithm	79
		7.1.4 Limitations	81
	7.2	AppSAT-Guided Removal (AGR) Attack	82
		7.2.1 Basic Idea	82
		7.2.2 Attack Algorithm	82
	7.3	Sensitization-Guided SAT (SGS) Attack............................	84
		7.3.1 Basic Idea	84
		7.3.2 Security Vulnerabilities of ATD	85
		7.3.3 Attack Algorithm	88
	7.4	Bypass Attack	89
		7.4.1 Basic Idea	89
		7.4.2 Attack Algorithm	90
	References		91

8   Post-SAT 2: Insertion of SAT-Unresolvable Structures 93
       8.1   Cyclic Logic Locking 93
              8.1.1   Basic Idea 93
              8.1.2   Non-reducible Cycles 94
              8.1.3   Cyclic Logic Locking Algorithm 95
              8.1.4   Security Analysis 96
       8.2   CycSAT 97
              8.2.1   Basic Idea 97
              8.2.2   Formulating NC Constraints 97
              8.2.3   Attack Algorithm 99
       8.3   ORF-Lock: One-Way Function-Based Logic Locking 99
              8.3.1   Basic Idea 99
              8.3.2   Methodology 100
              8.3.3   Security Analysis 100
       References 102

9   Post-SAT 3: Stripped-Functionality Logic Locking 103
       9.1   Motivation and Basic Concepts 103
              9.1.1   Motivation 103
              9.1.2   Variants of SFLL 104
       9.2   SFLL-HD^0: A Special Case of SFLL-HD 105
              9.2.1   Basic Idea 105
              9.2.2   Architecture 105
              9.2.3   Security Analysis 106
       9.3   SFLL-HD for Protecting Multiple Patterns 108
              9.3.1   Architecture 108
              9.3.2   Security Analysis 108
              9.3.3   Resilience Trade-Offs 110
       9.4   SFLL-Flex 111
              9.4.1   Architecture 112
              9.4.2   Optimization Framework 113
              9.4.3   Security Analysis 115
       References 117

10   Side-Channel Attacks 119
       10.1   Differential Power Analysis (DPA) Attack 119
              10.1.1   Basic Idea 119
              10.1.2   Preliminaries: The DPA Attack 120
              10.1.3   DPA Attack on Logic Locking 121
       10.2   Test-Data Mining (TDM) Attack 122
              10.2.1   Basic Idea 122
              10.2.2   TDM Attack Algorithm 123
              10.2.3   HackTest Attack on IC Camouflaging 125
       10.3   Hill Climbing Search Attack 127
              10.3.1   Basic Idea 127
              10.3.2   Attack Algorithm 127

10.4   De-synthesis Attack 128
        10.4.1   Basic Idea 128
        10.4.2   Attack Algorithm 129
References 130

11   Discussion 131
11.1   Revisiting the Attack/Defense Matrix 131
11.2   Challenges Faced by Logic Locking 133
11.3   Directions for Future Research 134
References 135

A   Background on VLSI Test 139
A.1   Manufacturing Test 139
A.2   Fault Models 139
A.3   Automatic Test Pattern Generation (ATPG) 140
A.4   Detection of a Stuck-at Fault 140
A.5   Scan-Based Testing 141

Acronyms

3PIP	Third-party intellectual property
AGR	AppSAT-guided removal
ATPG	Automatic test pattern generation
BA	Basic Anti-SAT
BEOL	Back end of line
DfTr	Design for trust
DPA	Differential power analysis
EPIC	Ending piracy of integrated circuits
FEOL	Front end of line
FLL	Fault analysis-based logic locking
FSM	Finite state machine
IC	Integrated circuit
IP	Intellectual property
LUT	Look-up table
OA	Obfuscated Anti-SAT
OC	Output corruptibility
OER	Output error rate
ORF	One-way random function
OSAT	Outsourced assembly and test
PPT	Probabilistic polynomial time
RLL	Random logic locking
SAT	Boolean satisfiability
SEM	Scanning electron microscope
SFLL	Stripped-functionality logic locking
SGS	Sensitization-guided SAT
SLL	Strong logic locking
SoC	System on chip
SPS	Signal probability skew
TDM	Test-data mining
TTLock	Tenacious and traceless logic locking

Chapter 1
The Need for Logic Locking

Abstract The first chapter of the book describes the need for logic locking and how it addresses the hardware security challenges faced by the IC design community. The chapter begins with a description of the globalized IC design flow and the associated security threats. This introduction is followed by a brief description of various design-for-trust countermeasures and their comparison with logic locking in terms of security properties. The chapter ends with a detailed description of logic locking and the associated terminology that will be used throughout this book.

Integrated circuits (ICs) are ubiquitous and an essential component in our lives today. ICs at the heart of electronic systems ranging from home appliances and smartphones to satellites and military equipment. ICs serve as the root-of-trust for these systems. The software computations can only be trustworthy if the underlying hardware is reliable and trustworthy [42]. The security of the hardware is, gradually, becoming as important as that of the software, in part due to the emergence of hardware attacks, such as the latest Spectre and Meltdown attacks on Intel processors [23, 29]. The primary reason for many hardware-based attacks, such as reverse engineering or IP piracy, is the profit-driven globalization of the IC design flow. This chapter focuses on logic locking, which is a well-known countermeasure against multiple hardware-based attacks. If we have to summarize logic locking in one sentence, it would be something along the following lines: Logic locking "locks" the functionality of a design until it is unlocked with a secret key. You might be wondering how a design is locked? The high-level answers to this and many other questions that you might have will be provided before the end of this chapter. The specific details of various logic locking algorithms can be found in the subsequent chapters of this book.

This chapter is organized as follows. Section 1.1 describes the globalized IC design flow followed by a description of the associated security threats in Sect. 1.2. Section 1.3 presents a summary of the existing design-for-trust (DfTr) countermeasures. Section 1.4 elaborates on the fundamental concepts of logic

© Springer Nature Switzerland AG 2020
M. Yasin et al., *Trustworthy Hardware Design: Combinational Logic Locking Techniques*, Analog Circuits and Signal Processing,
https://doi.org/10.1007/978-3-030-15334-2_1

locking and illustrates how it addresses various hardware security vulnerabilities. The same section also introduces various terms and definitions that are associated with logic locking and will be used throughout this book.

1.1 Globalization of the IC Design

As mentioned earlier, various security threats have emerged as a consequence of the globalization of the IC design flow. This section presents a brief description of the traditional and the globalized IC design flow. Note that only a rudimentary/high-level understanding of the IC design flow is required to understand the concepts presented in this chapter.

1.1.1 Traditional IC Design Flow

Figure 1.1 depicts the major steps involved in the traditional IC design flow. The overall process starts with a set of system specifications. The specifications are translated, using a series of design steps, collectively referred to as *logic design*, into a *netlist* consisting of interconnected logic gates, where each gate implements a Boolean function. The gate-level netlist then passes through another series of *physical design* steps, which map all the gates and interconnects in the netlist to an equivalent geometric representation of transistors and wires, referred to as a *layout*. The layout is then sent to a foundry, where a set of *masks* is constructed from it for use in photolithography of silicon *wafers*, leading to the *fabrication* of the chips/ICs. The fabricated ICs are then tested for manufacturing errors, and those that pass the *manufacturing test* are classified as functional ICs. These ICs are then distributed to the end-users for deployment in various electronic systems. Traditionally, leading semiconductor companies such as Intel and IBM used to follow a vertically-integrated business model, i.e., the companies would own foundries, and would design, fabricate, and test the ICs in-house, in trusted settings [42].

1.1.2 Globalized IC Design Flow

The evolving complexity of ICs, the ever-shrinking time-to-market, and the skyrocketing costs of building or maintaining a semiconductor foundry have propelled the globalization of IC design flow [9, 32]. Today, many semiconductor companies, e.g., Apple, operate on a *fabless* business model, i.e., they outsource the fabrication of ICs to offshore foundries, e.g., TSMC in Taiwan [34]. Apart from the fabrication of ICs,

Fig. 1.1 Major
design/processing steps
involved in the traditional IC
design flow. The difference
between the traditional and
the globalized design flow are
highlighted in red

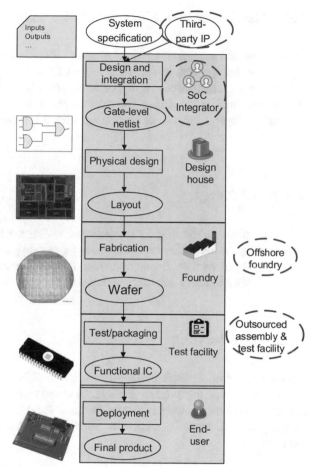

test, assembly, or packaging may also be outsourced to *outsourced assembly and test (OSAT)* companies, e.g., ASE [41] and Amkor [7]. Another common practice is for the design houses to purchase and incorporate third-party intellectual property *(3PIP)* cores in a design; this significantly reduces the design effort and helps satisfy the strict time-to-market constraints. Thus, multiple companies located all across the globe may be involved in the processing steps required for successful IC design and deployment. Figure 1.1 highlights the differences between the traditional and the globalized design flow using red circles. Note that the main entities handling outsourced services are the 3PIP vendor, SoC integrator, foundry, and OSAT. As the IC design flow is distributed worldwide, ICs are susceptible to new kinds of threats, which are the focus of the next section.

1.2   The Emergence of Hardware Security Vulnerabilities

In a globalized and distributed IC supply chain, several (potentially untrusted) agents may have access to the valuable soft IP or the physical IC. The increased access to critical assets opens up opportunities for rogue elements to exploit valuable information and/or jeopardize the trust in the IC design flow [15, 16, 42, 43]. Figure 1.2 depicts the threats associated with the main entities in the IC design flow.

1.2.1   IP Piracy

IP piracy refers to the illegal or unlicensed use of the intellectual property (e.g., netlist or layout files). Similar to the case of software piracy, an attacker in a design house, can steal valuable IP cores and sell them as genuine. The yearly losses to the semiconductor industry due to IP infringement exceed $4 Billion [44].

1.2.2   Overbuilding

Piracy may manifest itself in the form of overbuilding at an untrusted foundry. A rogue foundry can overproduce extra ICs and sell them illegally. Building extra ICs from the same set of masks increases the costs for the foundry only marginally; the foundry can sell those ICs at a price cheaper than that offered by the original IC company [42].

1.2.3   Hardware Trojans

Malicious agents in a design house or a foundry can insert small, hard-to-detect circuitry in an IC that can reveal secret information or can cause service disruption at a specific moment during the IC operation [21]. As the footprint for a Trojan may be very small, it becomes hard to verify/test for the presence of hardware Trojans in a circuit, especially when there is no golden (Trojan-free) reference available to compare against. The suspected sources of Trojans are 3PIP used in the design or mask alteration during fabrication [8].

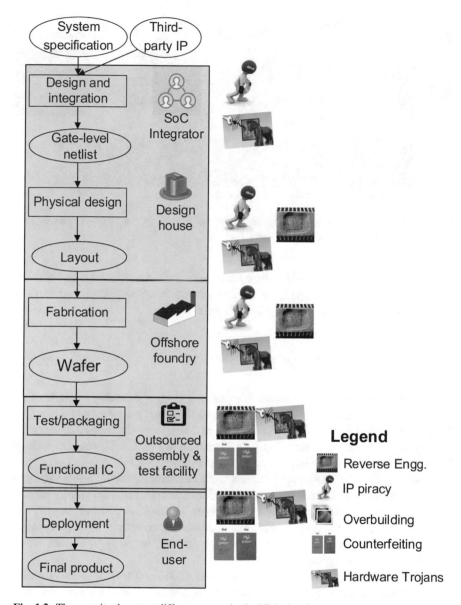

Fig. 1.2 The security threats at different stages in the IC design flow

1.2.4   Counterfeiting

Counterfeit ICs are replicas of the genuine ICs that are fraudulently made to appear almost-identical to the genuine ICs [14]. Counterfeit ICs include out-of-spec, remarked, and recycled ones, typically salvaged from discarded electronic appliances. More than 5% of all commercial ICs are counterfeits [14, 16]. Counterfeit

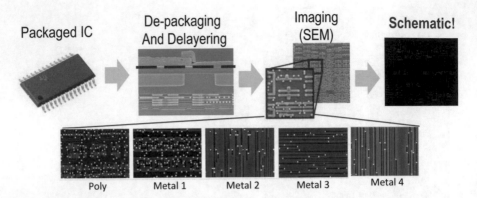

Fig. 1.3 Steps involved in reverse engineering of an IC. The IC involved is a smart card undergoing reverse engineering at Texplained [45]

ICs pose serious reliability and security concerns for the semiconductor industry. According to a 2012 IHS report, the top five counterfeited semiconductor components represent a revenue risk of up to $169 Billion [40].

1.2.5   Reverse Engineering

Reverse engineering of an IC refers to extracting design/technology details of an IC using imaging techniques. It involves several steps that include: depackaging an IC, delayering and imaging individual layers, and analyzing the collected images to identify netlist/schematic details [48], as illustrated in Fig. 1.3. Companies such as Chipworks provide reverse engineering as a service [13].

Apart from use in benign applications such as checking for IP rights infringement, reverse engineering can also be an enabler for hardware-based attacks such as IP piracy, counterfeiting, or insertion of hardware Trojans. An attacker may exploit reverse engineering to identify stealthy locations for inserting Trojans. Reverse engineering can be used in conjunction with probing methods to extract secret information such as cryptographic keys [14].

In summary, a globalized IC design flow has opened up new avenues of attack for malicious and untrusted entities. There is an imminent need for solutions that protect ICs and electronic systems against these threats.

1.3   Design-for-Trust (DfTr) Solutions

A number of design-for-trust techniques have been proposed to thwart IP piracy, reverse engineering, and other hardware-based attacks. These techniques include watermarking [19], fingerprinting [10], metering [24], camouflaging [38], split manufacturing [18], and logic locking [11, 12, 27, 35, 36, 39, 39, 43]. A description of these countermeasures is presented as follows.

1.3.1 Watermarking and Fingerprinting

In watermarking, the designer's signature, e.g., a secret design constraint, is embedded into the design; whereas in fingerprinting, the end-user's signature is embedded along with the designer's signature to help track the source of piracy. Watermarking and fingerprinting are *passive* techniques that only help detect piracy but cannot prevent it from occurring [10, 19]. Both techniques can be employed during either logic or physical design stage.

1.3.2 Camouflaging

Camouflaging is a countermeasure that aims at preventing reverse engineering carried out by end-users. It replaces selected gates in a design with their camouflaged counterparts; the camouflaged gates look alike from the top but can implement one of many functions. Upon imaging using an optical microscope or even an scanning electron microscope (SEM), the true functionality of the camouflaged gates remains unknown. Thus, a reverse-engineering attacker faces the problem of resolving the functionality of the camouflaged gates through additional efforts [3, 28, 38, 47, 52].

Camouflaging can be performed by using dummy contacts [47], filler cells [4], or diffusion programmable standard cells [6, 46]. Camouflaging using dummy contacts is illustrated in Fig. 1.4, where each camouflaged cell can implement one of two functions: NAND or NOR. Camouflaging is a layout-level technique that is carried out after the initial physical design of an IC. It requires the support of a trusted foundry for fabrication of the camouflaged gates. Syphermedia offers a library of camouflaged standard cells that have typically been.

1.3.2.1 Split Manufacturing

In split manufacturing, the design is split into two parts, corresponding to back-end-of-the-line (BEOL) and front-end-of-the-line (FEOL) metal layers, that are manufactured in separate foundries and ultimately stacked together [18]. Split manufacturing is also a layout-level countermeasure; the layout split is effected after the initial physical design of an IC. It can prevent against piracy only by an untrusted foundry, but not by an end-user.

Fig. 1.4 Layout of typical
2-input (**a**) NAND and (**b**)
NOR gates. The metal layers
look different from the top,
and it is easy to distinguish by
visual inspection.
Camouflaged layout of
2-input (**c**) NAND and (**d**)
NOR gates [38]. The metal
layers are identical, and the
two gates cannot be
distinguished from the top
view

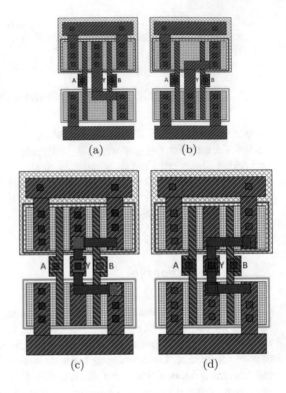

1.3.3 Metering

Metering is a set of protocols and techniques, either passive or active, which assign
a unique ID to each manufactured IC and help in tracking of individual ICs post-
manufacturing [1]. While the passive metering techniques aim at only identifying
piracy, active metering techniques allow the IC owner to monitor and control the IC
behavior during in-field operation. A comprehensive review of hardware metering
techniques can be found in [24].

1.3.4 Logic Locking: An Overview

Logic locking inserts additional logic into a design that locks it with a secret
key. Only upon unlocking/activation with the correct key, the design is functional.
Figure 1.5 illustrates that the output of a *locked* circuit is a function of both the
primary inputs and the additional key inputs. Upon applying incorrect key to the
design, an incorrect output is obtained. In this section, we restrict the discussion to
a high level comparison of logic locking with the other DfTr techniques. A detailed
description of logic locking and how it offers protection against specific threats is

Fig. 1.5 An abstract representation of a logic locked design. Only on applying the secret key, the design produces the correct output; otherwise, an incorrect output is produced

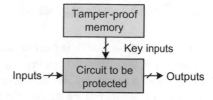

Table 1.1 Protection offered by DfTr techniques against IP piracy, overbuilding, and reverse engineering attacks, conducted by various untrusted entities involved in the IC design flow

DfTr technique	SoC integrator	Foundry	Test facility	End-user
Watermarking [19, 20, 22, 33]	✗	✗	✗	✗
Camouflaging [2, 3, 28, 30, 38, 47, 50]	✗	✗	✗	✓
Split manufacturing [17, 18]	✗	✓	✗	✗
Metering (passive) [1, 25, 26]	✗	✗	✓	✓
Logic locking [37, 39, 43, 51]	✓	✓	✓	✓

✓ denotes that a technique can protect piracy conducted by an untrusted entity
✗ denotes the inability to defend against attacks

presented in Sect. 1.4. Logic locking offers the most versatile protection of all DfTr techniques. It assumes only the design house to be trusted and can protect against piracy, overbuilding, and reverse engineering attacks by rogue entities present at any subsequent stage in the IC design flow. Compared to metering, logic locking locks all ICs with a common key. It may be integrated with metering techniques to assign a unique key to each fabricated IC.

1.3.5 Logic Locking vs. Other DfTr Techniques

It is clear from the above discussion that the DfTr techniques differ mainly in the threat model and security objectives. Threat model specifies the entities that are trusted (or untrusted) and the assets each entity has access to. Table 1.1 presents a comparison of the trust models for different DfTr techniques. As already highlighted, watermarking and fingerprinting are passive techniques that can only help detect piracy/theft once it has occurred; they do not prevent piracy from occurring. Camouflaging and split manufacturing are layout-level techniques and such can offer protection only after this stage. Camouflaging requires the support of a trusted foundry for manufacturing the camouflaged gates; accordingly, it offers protection only against untrusted end-users. Split manufacturing aims at protecting only against an untrusted foundry. While camouflaging and split manufacturing prevent against piracy by only a single entity (end-user and foundry, respectively), logic locking can protect against rogue agents involved anywhere in the design flow, except for the trusted design house. In contrast to camouflaging, logic locking does not require the support of the foundry. Furthermore, logic locking does not need a trusted BEOL foundry, which is required for split manufacturing.

1.4   Logic Locking: Definitions and Terminology

Logic locking is a DfTr technique that protects against reverse engineering-based piracy of ICs by inserting additional logic into a circuit and locking the original design with a secret key. In addition to the original inputs, a locked circuit has *key inputs* that are driven by an on-chip tamper-proof memory [31, 49], as illustrated in Fig. 1.5. The additional logic may consist of either combinational logic such as XOR gates [37, 39, 43], or sequential logic such as look-up tables (LUTs) [5]. Only upon application of the correct key value, the design is functional and produces the correct output; otherwise, its output differs from that of the original design. The security properties of a logic locking technique are dictated by the protection logic employed. This will be elaborated in Chap. 2 and multiple subsequent chapters.

1.4.1   IC Design Flow with Logic Locking

Figure 1.6 presents the IC design flow incorporating logic locking. The original netlist is locked using a secret key value, known only to the IP owner. The locked netlist passes through the untrusted design stages including an untrusted foundry and the outsourced test facility, as highlighted in Fig. 1.6. Without the knowledge of the correct key, an attacker cannot recover the correct functionality of the design. The IP owner activates a fabricated IC by loading the secret key to the tamper-proof memory, in trusted settings.

Example Figure 1.7a shows an example (majority circuit) netlist and Fig. 1.7b shows its locked version, with three additional XOR/XNOR key gates. One of the inputs of each key gate is driven by a wire in the original design, while the other input, referred to as *key input*, is driven by a key bit stored in a tamper-proof memory [49]. A locked IC (or a locked netlist) will not generate correct output unless it is activated using the correct key. Consider the locked circuit in Fig. 1.7b. When the correct key value 110 is loaded to the memory, all the key gates in the circuit behave as buffers and produce the correct output, for all input patterns. For example, the locked circuit output $Y = 0$ for the input pattern 000. However, when an incorrect key value is applied, certain key gates will behave as inverters, leading to injection of an error in the circuit. Let us consider 010 as an incorrect key value and again, apply the input pattern 000. The key gate K1 behaves as an inverter, leading to an incorrect output $Y = 1$. Table 1.2 presents the output of the locked circuit for all key and primary input combinations. It can be observed that the error injected into the circuit is not uniform across the input patterns. For certain input patterns such as 010, all key values except the correct key $k6$ produce an incorrect output. For other input patterns such as 011, no error is injected at all.

Fig. 1.6 Locking and activation of an IC in the context of the IC design flow. The red regions represent the untrusted entities; the green regions represent the trusted ones

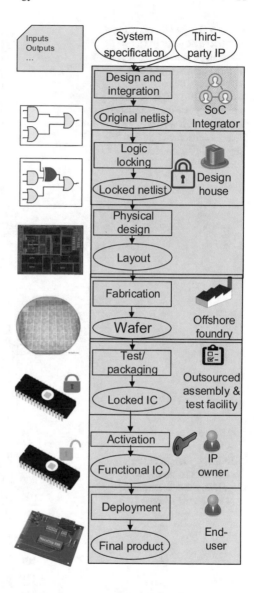

There exists a one-to-one mapping between the key-gate type and the key-gate value in Fig. 1.7b, the correct key value for an XNOR key gate is 1, and that for an XOR key gate is 0. This suggests that an attacker may infer the secret key value directly from the types of the key gate. However, as illustrated in Fig. 1.7c, such simple attacks can be easily thwarted by pushing inverters around in the netlist, effectively decoupling the values of the key gates from their types.

Fig. 1.7 Logic locking using XOR/XNOR gates [43]. (**a**) An example circuit: majority of three inputs. (**b**) Circuit locked using three XOR/XNOR key gates. The correct key is value is 110. (**c**) Locked circuit with the inverters absorbed into the key gates. The correct key value is still 110

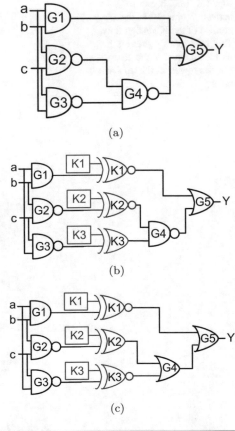

(a)

(b)

(c)

Table 1.2 Output of the locked circuit, shown in Fig. 1.7a, for different input and key combinations

abc	Y	k0	k1	k2	k3	k4	k5	k6	k7
000	0	1	1	1	1	1	1	0	1
001	0	1	1	1	1	1	1	0	1
010	0	1	1	1	1	1	1	0	1
011	1	1	1	1	1	0	1	1	1
100	0	1	1	1	1	1	1	0	1
101	1	1	1	1	1	1	1	1	0
110	1	1	1	0	1	1	1	1	1
111	1	1	0	1	1	1	1	1	1

1.4.2   Terminology

Original and Locked Netlist   The original netlist F is a Boolean function $F : I \rightarrow O$, where $I = \{0, 1\}^n$ and $O = \{0, 1\}^m$. The locked netlist is a Boolean function $L : I \times K \rightarrow O$, where $K = \{0, 1\}^k$, where k is *key size* and denotes the number of key bits. Upon activation using the secret key k_s, $L(i, k_s) = F(i), \forall i \in I$, i.e.

locked netlist output is identical with that of the original netlist F. Note that the term circuit and netlist may be used alternatively in this book, without any loss of generality.

Security of Logic Locking A logic locking technique is considered secure if the effort required by an attacker to determine the correct key value k_s, or retrieve the original circuit functionality F, is exponential in the key size $|k|$, i.e., $\mathcal{O}(2^{|k|})$.

1.4.3 Protection Against Hardware-Based Attacks

In this section, we describe how logic locking protects against various hardware-based attacks.

- **IP piracy and reverse engineering.** An adversary can either steal the locked netlist or obtain it by the laborious process of reverse engineering an IC, which can take months to complete for typical ICs. However, without the knowledge of the correct key, the extracted netlist will not fully correspond to the IC and will be of little use to the attacker. As already mentioned, in XOR-based logic locking, key gates replace either functional buffers or inverters (key values are determined accordingly), introducing ambiguity in design reconstruction. Note that logic locking hides some part of the functionality in the secret key bits; the structural information obtained from hardware implementation reveals only the remaining part of the functionality.
- **Overbuilding.** A malevolent foundry (or a rogue employee in the foundry) can produce extra ICs to sell them illegally at lower prices. However, without the secret key, the fabricated ICs cannot be activated and remain non-functional.
- **Hardware Trojans.** Logic locking can help prevent the insertion of Trojans in a circuit by judiciously altering the transition probabilities of the signals in a circuit, in a manner unknown to the attacker, and making it harder for the attacker to identify stealthy candidates for inserting Trojans.
- **Counterfeiting.** Logic locking aims primarily at protecting against piracy. As such, it may not prevent all forms of counterfeiting. However, it can hamper selected instances of counterfeiting, such as cloning, which relies heavily on reverse engineering [16].

1.5 Takeaway Points

This chapter elaborated on multiple hardware security threats and how they can be thwarted using different DfTr techniques. It also highlighted the advantages of logic locking over other DfTr techniques and described how logic locking helps thwart various hardware-based attacks. Chapter 2 will present a comprehensive history of logic locking techniques as well as attacks that have been launched against logic locking.

References

1. Alkabani Y, Koushanfar F (2007) Active hardware metering for intellectual property protection and security. In: USENIX security, pp 291–306
2. Baukus J, Chow L, Cocchi R, PO, Wang B (2012) Building block for a secure CMOS logic cell library. US Patent no. 8111089
3. Baukus J, Chow L, Cocchi R, Wang B (2012) Method and apparatus for camouflaging a standard cell based integrated circuit with micro circuits and post processing. US Patent no. 20120139582
4. Baukus JP, Chow LW, Cocchi RP, Ouyang P, Wang BJ (2012) Camouflaging a standard cell based integrated circuit. US Patent no 8151235
5. Baumgarten A, Tyagi A, Zambreno J (2010) Preventing IC piracy using reconfigurable logic barriers. IEEE Des Test Comput 27(1):66–75
6. Becker GT, Regazzoni F, Paar C, Burleson WP (2013) Stealthy dopant-level hardware trojans. In: Cryptographic hardware and embedded systems. Springer, Berlin, pp 197–214
7. Berry M, John G (2014) Outsourcing Test – What are the most valuable engagement periods? http://www.amkor.com/go/outsourcing-test [May 16, 2016]
8. Bhunia S, Hsiao MS, Banga M, Narasimhan S (2014) Hardware trojan attacks: threat analysis and countermeasures. Proc IEEE 102(8):1229–1247
9. Board DS (2005) Defense Science Board (DSB) study on High Performance Microchip Supply. www.acq.osd.mil/dsb/reports/ADA435563.pdf [March 16, 2015]
10. Caldwell AE, Choi HJ, Kahng AB, Mantik S, Potkonjak M, Qu G, Wong JL (1999) Effective iterative techniques for fingerprinting design IP. In: IEEE/ACM design automation conference, pp 843–848
11. Chakraborty RS, Bhunia S (2009) HARPOON: an obfuscation-based SoC design methodology for hardware protection. IEEE Trans Comput Aided Des Integr Circuits Syst 28(10):1493–1502
12. Chakraborty RS, Bhunia S (2009) Security against hardware trojan through a novel application of design obfuscation. In: IEEE/ACM international conference on computer-aided design, pp 113–116
13. Chipworks (2016) Reverse engineering software. http://www.chipworks.com/en/technical-competitive-analysis/resources/reerse-engineering-software
14. Goertzel KM, Hamilton B (2013) Integrated circuit security threats and hardware assurance countermeasures. CrossTalk, p 33
15. Guin U, DiMase D, Tehranipoor M (2007) Counterfeit integrated circuits: detection, avoidance, and the challenges ahead. J Electron Test 30(1):9–23
16. Guin U, Huang K, DiMase D, Carulli JM, Tehranipoor M, Makris Y (2014) Counterfeit integrated circuits: a rising threat in the global semiconductor supply chain. Proc IEEE 102(8):1207–1228
17. Imeson F, Emtenan A, Garg S, Tripunitara MV (2013) Securing computer hardware using 3D integrated circuit (IC) technology and split manufacturing for obfuscation. In: USENIX conference on security, pp 495–510
18. Jarvis R, McIntyre M (2007) Split manufacturing method for advanced semiconductor circuits. US Patent 7,195,931
19. Kahng A, Lach J, Mangione-Smith WH, Mantik S, Markov I, Potkonjak M, Tucker P, Wang H, Wolfe G (1998) Watermarking techniques for intellectual property protection. In: IEEE/ACM design automation conference, pp 776–781
20. Kahng A, Mantik S, Markov I, Potkonjak M, Tucker P, Wang H, Wolfe G (1998) Robust IP watermarking methodologies for physical design. In: Design automation conference, pp 782–787
21. Karri R, Rajendran J, Rosenfeld K, Tehranipoor M (2010) Trustworthy hardware: identifying and classifying hardware trojans. Computer 43(10):39–46
22. Kirovski D, Potkonjak M (2003) Local watermarks: methodology and application to behavioral synthesis. IEEE Trans Comput Aided Des Integr Circuits Syst 22(9):1277–1283

23. Kocher P, Genkin D, Gruss D, Haas W, Hamburg M, Lipp M, Mangard S, Prescher T, Schwarz M, Yarom Y (2018) Spectre attacks: exploiting speculative execution. arXiv:1801.01203
24. Koushanfar F (2011) Integrated circuits metering for piracy protection and digital rights management: an overview. In: Great lakes symposium on VLSI, pp 449–454
25. Koushanfar F (2012) Provably secure active IC metering techniques for piracy avoidance and digital rights management. IEEE Trans Inf Forensics Secur 7(1):51–63
26. Koushanfar F, Qu G (2001) Hardware metering. In: IEEE/ACM design automation conference, pp 490–493
27. Lee YW, Touba N (2015) Improving logic obfuscation via logic cone analysis. In: Latin-American test symposium, pp 1–6
28. Li M, Shamsi K, Meade T, Zhao Z, Yu B, Jin Y, Pan D (2016) Provably secure camouflaging strategy for IC protection. In: IEEE/ACM international conference on computer-aided design, pp 28:1–28:8
29. Lipp M, Schwarz M, Gruss D, Prescher T, Haas W, Mangard S, Kocher P, Genkin D, Yarom Y, Hamburg M (2018) Meltdown. arXiv:1801.01207
30. Massad M, Garg S, Tripunitara M (2015) Integrated circuit (IC) decamouflaging: reverse engineering camouflaged ICs within minutes. In: Network and distributed system security symposium
31. Maxim Integrated (2010) Deepcover security manager for low-voltage operation with 1kb secure memory and programmable tamper hierarchy. https://www.maximintegrated.com/en/products/embedded-security/security-managers/DS3660.html
32. Mazurek J (1999) Making microchips: policy, globalization, and economic restructuring in the semiconductor industry. MIT Press, Cambridge
33. Oliveira A (1999) Robust techniques for watermarking sequential circuit designs. In: IEEE/ACM design automation conference, pp 837–842
34. Patterson A (2017) Apple talks about sole sourcing from TSMC. http://www.eetimes.com/document.asp?doc_id=1332496 [Feb 16, 2018]
35. Plaza S, Markov I (2015) Solving the third-shift problem in IC piracy with test-aware logic locking. IEEE Trans Comput Aided Des Integr Circuits Syst 34(6):961–971
36. Rajendran J, Pino Y, Sinanoglu O, Karri R (2012) Logic encryption: a fault analysis perspective. In: Design, automation and test in Europe, pp 953–958
37. Rajendran J, Pino Y, Sinanoglu O, Karri R (2012) Security analysis of logic obfuscation. In: IEEE/ACM design automation conference, pp 83–89
38. Rajendran J, Sam M, Sinanoglu O, Karri R (2013) Security analysis of integrated circuit camouflaging. In: ACM/SIGSAC conference on computer & communications security, pp 709–720
39. Rajendran J, Zhang H, Zhang C, Rose G, Pino Y, Sinanoglu O, Karri R (2015) Fault analysis-based logic encryption. IEEE Trans Comput 64(2):410–424
40. Release IP (2012) Top 5 most counterfeited parts represent a $ 169 billion potential challenge for global semiconductor market. http://press.ihs.com/press-release/design-supply-chain/top-5-most-counterfeited-parts-represent-169-billion-potential-cha [June 30, 2015]
41. Room AP (2014) Silicon Laboratories and ASE announce milestone shipment of 10 million tested integrated circuits. http://www.aseglobal.com/en/News/PressRoomDetail.aspx?ID=45 [May 16, 2016]
42. Rostami M, Koushanfar F, Karri R (2014) A primer on hardware security: models, methods, and metrics. Proc IEEE 102(8):1283–1295
43. Roy J, Koushanfar F, Markov IL (2010) Ending piracy of integrated circuits. IEEE Comput 43(10):30–38
44. SEMI (2008) Innovation is at risk losses of up to $4 billion annually due to IP infringement. www.semi.org/en/Issues/IntellectualProperty/ssLINK/P043785, [June 10, 2015]
45. Shen Y, Zhou H (2015) Advanced IC reverse engineering techniques: in depth analysis of a modern smart card. Black Hat 2015. https://www.blackhat.com/docs/us-15/materials/us-15-Thomas-Advanced-IC-Reverse-Engineering-Techniques-In-Depth-Analysis-Of-A-Modern-Smart-Card.pdf

46. Shiozaki M, Hori R, Fujino T (2014) Diffusion programmable device: the device to prevent reverse engineering. IACR Cryptology ePrint Archive, p 109
47. SypherMedia (2015) SypherMedia Library Circuit Camouflage Technology. http://www.smi.tv/solutions.htm [Aug 10, 2015]
48. Torrance R, James D (2011) The state-of-the-art in semiconductor reverse engineering. In: IEEE/ACM design automation conference, pp 333–338
49. Tuyls P, Schrijen G, Škorić B, van Geloven J, Verhaegh N, Wolters R (2006) Read-proof hardware from protective coatings. In: Goubin L, Matsui M (eds) International conference on cryptographic hardware and embedded systems, pp 369–383
50. Vijayakumar A, Patil V, Holcomb D, Paar C, Kundu S (2017) Physical design obfuscation of hardware: a comprehensive investigation of device and logic-level techniques. IEEE Trans Inf Forensics Secur 12(1):64–77
51. Xie Y, Srivastava A (2016) Mitigating SAT attack on logic locking. In: International conference on cryptographic hardware and embedded systems, pp 127–146
52. Yasin M, Mazumdar B, Sinanoglu O, Rajendran J (2016) CamoPerturb: secure IC camouflaging for minterm protection. In: IEEE/ACM international conference on computer-aided design, pp 29:1–29:8

Chapter 2
A Brief History of Logic Locking

Abstract This chapter presents a comprehensive history of logic locking defenses and attacks. A classification of logic locking techniques as well as attacks is provided. The logic locking defenses are divided into classes: pre-SAT and post-SAT techniques. Four classes of attacks: algorithmic, approximate, structural, and side-channel, are introduced. The chapter emphasizes the relationship between different logic locking techniques. A timeline of the prominent logic locking attacks and defenses is also presented.

Since the inception of logic locking in 2008 [16], it has received significant interest from the research community. Over the last decade, a number of logic locking techniques as well as attacks have emerged. This chapter presents a high-level introduction to the major developments in logic locking. In addition to offering an overview of different logic locking attacks and defenses, the chapter also highlights the relationships between different attack and defense algorithms. Section 2.1 provides a summary of the milestones in logic locking research. Section 2.2 introduces a classification of logic locking defenses and attacks. Section 2.3 presents a brief overview of the existing logic locking defenses. Section 2.4 summarizes the existing attacks on logic locking. Section 2.5 elaborates on the resilience of each defense against different attack algorithms using an attack-defense matrix. Section 2.6 presents a summary of different metrics that can be used to evaluate the effectiveness of a logic locking technique.

2.1 Milestones in Logic Locking

This section focuses on the milestones in logic locking research. We introduce the first logic locking defense, describe the first attack against it, and also discuss the most powerful attack on logic locking. While detailed descriptions of these attacks and defense techniques are presented in the relevant chapters, only an

© Springer Nature Switzerland AG 2020

M. Yasin et al., *Trustworthy Hardware Design: Combinational Logic
Locking Techniques*, Analog Circuits and Signal Processing,
https://doi.org/10.1007/978-3-030-15334-2_2

abstract overview is presented here so that the reader becomes familiar with the key advancements in the field of logic locking.

2.1.1 The First Defense

The first logic locking technique, introduced in 2008, is random logic locking (RLL) [16]. As the name indicates RLL inserts XOR/XNOR key gates at random locations in a netlist; only upon supplying the correct key to a locked chip, it becomes functional. To put things in a historical perspective, it must be mentioned that RLL was introduced in the context of IC metering. The overall IC metering framework EPIC, which abbreviates "Ending Piracy of Integrated Circuits", not only locks the design with a common locking key CK but also provides additional circuitry for generating unique key values for each manufactured IC. EPIC also makes use of public-key cryptography and enables remote activation of a chip. A designer can communicate remotely with the chip fabricated at an untrusted foundry, and can securely load the key CK onto the chip. While an interested reader can refer to [16, 17] for details of the EPIC protocol, it suffices to say that at the core of EPIC is RLL that renders the functionality of the fabricated chip dependent on the key inputs. We discuss RLL in detail in Sect. 3.1.

2.1.2 The First Threat Model and Attack

While RLL was the first defense logic locking defense and inspired further research on IC/IP piracy [1, 13], it was discovered by Rajendran et al. [14] in 2012 that the individual key bits in the common key CK that activates a functional IC may be observed on the primary outputs of IC. The new threat model introduced by Rajendran et al. assumes that the attacker has access to two critical assets, (1) a reverse-engineered netlist, and (2) a functional IC. This threat model has been adopted by all subsequent logic locking attacks and defenses. Their proposed *sensitization* attack, which follows the aforementioned threat model proceeds as follows. By analyzing the locked netlist, the attack computes attack patterns. By applying these patterns to a functional IC, specific key bits may be sensitized to primary outputs of the IC, thus leaking the secret key. In RLL, most key bits do not interfere in each other's path to the primary outputs, and can be targeted on an individual basis [13].

As a countermeasure against the sensitization attack, strong logic locking (SLL) was introduced that inserts key-gates in a way that they protect one another [14] (see Sect. 3.4.3 for details).

2.1.3   The Most Powerful Attack

Apart from RLL and SLL, other important research efforts on logic locking include fault-analysis based logic locking (FLL) [13] and look-up table based locking [1]. However, in 2015, Pramod et al. [21] developed a powerful attack on logic locking that could break all the defense techniques existing then. The attack utilizes a Boolean satisfiability (SAT) formulation to encode the problem of finding the logic locking key and is commonly referred to as the *SAT attack*. The attack uses specific *distinguishing input patterns (DIPs)* to refine the key search space iteratively. The SAT attack has drastically changed the focus of logic locking research; most recent research efforts aim at developing effective countermeasures against the SAT attack [10, 22, 25, 30]. Chapter 4 discusses the SAT attack in further details.

2.2   Classification of Attacks and Defenses

2.2.1   Classifying Logic Locking Techniques

There can be multiple ways to classify the existing logic locking techniques. For instance, based on the type of logic elements constituting the protection circuitry, logic locking techniques can broadly be classified as either combinational or sequential. Whereas the combinational logic locking techniques insert XOR/XNOR gates [14, 15, 17], AND/OR gates [4], or multiplexers [12, 15] into a circuit, the sequential logic locking techniques insert look-up tables (LUTs) [1, 6], or finite state machines (FSMs) [2, 8]. Most of the recent research efforts focus on combinational logic locking techniques, as the tools and methods employed for the security analysis of the combinational techniques are mature and easily available. More importantly, the sequential circuits can often be treated as combinational by leveraging the test infrastructure on a chip, with scan chains being the most commonly used test access mechanism [12, 17]. This book focuses on combinational logic locking.

We classify the combinational logic locking techniques into major classes, (1) Pre-SAT and (2) Post-SAT logic locking. This classification is relative to the inception of the SAT attack [21]. Pre-SAT logic locking techniques are those that have been developed prior to the SAT attack and remain susceptible to the attack. Post-SAT techniques are those that have been developed after the SAT attack with the objective of thwarting the attack.

1. **Pre-SAT logic locking**. The techniques in the pre-SAT era focused on developing algorithms for selecting the key gate locations. These basic logic locking techniques include random logic locking (RLL) [17], fault analysis based logic locking (FLL) [15], and strong logic locking (SLL) [27].
2. **Post-SAT logic locking**. The research on logic locking took a whole new turn upon the inception of the SAT attack [21]; the attack was able to break all tradi-

tional logic locking techniques. Recent research efforts, such as SARLock [25], Anti-SAT [22], TTLock [31], and SFLL [30], focus on thwarting the SAT attack.

An introductory overview of various logic locking techniques is presented in Sect. 2.3; further details about each technique are furnished in the relevant chapters.

2.2.2   Classifying Attacks on Logic Locking

Over the years, a number of attacks have been developed against logic locking techniques. We divide these attacks into the following four categories: algorithmic attacks, approximate attacks, structural/removal attacks, and side-channel attacks.

1. **Algorithmic attacks**. These attacks exploit the algorithmic weaknesses of a logic locking algorithm to extract the secret key. Since the secret key renders the functionality of the locked netlist "exactly" equivalent to that of the functional IC, these attacks may also be referred to as "exact" attacks. Examples are the sensitization attack [14], the SAT attack [21], and the circuit partitioning (CP) attack [9].
2. **Approximate attacks**. Contrary to the exact attacks, the approximate attacks extract a netlist that is approximately the same as the original netlist, i.e., the netlist may produce an incorrect output for only a few input patterns. This category of attacks includes AppSAT [19] and Double-DIP [20]. These attacks require lesser computational effort compared to the exact attacks.
3. **Structural attacks**. The fundamental principle of the structural/removal attacks is to bypass and/or remove the protection logic and isolate the functionally correct netlist. Examples include the signal probability skew attack [26], the AppSAT guided removal attack [29], and the Bypass attack [23].
4. **Side-channel attacks**. Side-channel attacks exploit the covert physical channels, such as power and timing, to leak the secret information [7]. It has been demonstrated that certain logic locking techniques may be susceptible to differential power analysis attack [24]. Test data is another side-channel that has been used to compromise the security of logic locking [12, 28]. A recent attack that can be categorized as a side-channel attack is the desynthesis attack [11]; the attack exploits the security vulnerabilities associated with logic synthesis.

A brief introduction to the aforementioned attacks is provided in Sect. 2.4; further details are presented in the subsequent chapters.

2.2.3   A Timeline of Logic Locking

Figure 2.1 presents a timeline view of prominent logic locking attacks and defenses. The defenses are presented above the timeline and the attacks below it. The color of each dot represents the class of an attack or a defense. Following the classification

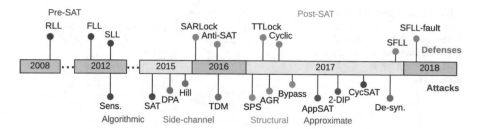

Fig. 2.1 A timeline view of logic locking attacks and defenses

mentioned above, there are two classes of defenses and four classes of attacks. It can be observed that the logic locking research has taken off 2015 on-wards with the emergence of multiple attacks and defenses. We believe that presenting this timeline earlier in the chapter puts things in a better perspective for the reader. Towards the end of this chapter, we also present an attack-defense matrix that elaborates on the connection between different attacks and countermeasures.

2.3 An Overview of Existing Defenses

2.3.1 Pre-SAT Logic Locking

All pre-SAT logic locking techniques insert either XOR/XNOR key gates or MUX key gates at selected locations in a netlist. These techniques differ mainly in the key gate location selection algorithm. As already mentioned, there exist three basic logic locking techniques: RLL [17], FLL [1, 15], SLL [14, 27]. Other gate selection algorithms such as [3–5] can be considered as variants of the three basic techniques. Here, we only present a brief description of the basic logic locking techniques. The detailed algorithms and illustrative examples are presented in Chap. 3.

RLL As already pointed out in the discussion of logic locking milestones, RLL inserts XOR/XNOR key gates at random locations in a netlist [17]. Figure 2.2a presents a netlist locked using RLL. RLL remains vulnerable to sensitization and SAT attacks [14, 21].

FLL A limitation of RLL is that even incorrect key values may lead to a correct circuit output for a large fraction of input patterns, enabling the black-box usage of a chip. FLL aims at preventing black-box usage of a locked chip [15]. It ensures that the maximum error is observed at the circuit output upon the application of incorrect key values. The *output corruption* is measured in terms of the percentage Hamming distance between the correct output and the incorrect output, obtained upon applying incorrect keys.

In FLL, the key gates are inserted at the most *influential* locations in the circuit, i.e., the locations that exhibit the highest impact on the circuit output upon the

Fig. 2.2 Key gate insertion based on basic logic locking techniques. (**a**) Random [17], (**b**) fault analysis-based [15], and (**c**) strong logic locking [27]

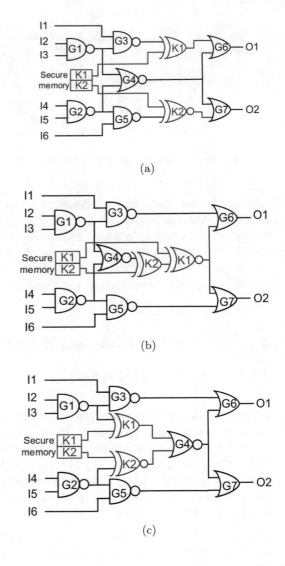

application of incorrect key values. Figure 2.2b shows a netlist locked using FLL. FLL is described in more detail in Sect. 3.2.

SLL   As pointed earlier, SLL aims at thwarting the sensitization attack by inserting key gates in a way that maximizes the interference among the key-gates and prevents the sensitization of the key bits on an individual basis. With increased interference among the key-gates, an attacker is forced to brute-force an exponentially growing number of key combinations [14].

Consider the netlist in Fig. 2.2c with two key-gates, K1 and K2, that are inserted using SLL. It can be seen that K1 and K2 interfere with each other's path to the primary outputs. It is not possible for an attacker to sensitize either K1 or K2 to a primary output on an individual basis. Refer to Sect. 3.4.3 for further details on SLL.

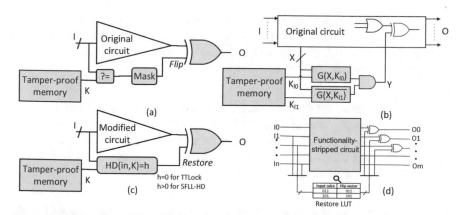

Fig. 2.3 SAT attack resilient logic locking techniques: (**a**) SARLock [25], (**b**) Anti-SAT [22], (**c**) SFLL-HD [30]/TTLock [31], and (**d**) SFLL-flex [30]

2.3.2 Post-SAT Logic Locking

The techniques developed recently to mitigate the SAT attack include SAR-Lock [25], Anti-SAT [22], ATD [10], TTLock [31], and SFLL [30]. Figure 2.3 illustrates the recent SAT attack resilient logic locking techniques. The common denominator for most of these techniques is a point function, which injects error for selectively into a circuit in a way that thwarts the SAT attack. A point function is a Boolean function that produces the output value 1 at exactly one point. Example implementations include AND gates and password checkers. Point functions help control the amount of error injected into a circuit upon the application of incorrect key values.

SARLock As shown in Fig. 2.3a, SARLock protection circuitry comprises a comparator and a mask block that are integrated with the original circuit [25]. For the correct key value, no error is injected into the circuit, and the correct output is retained. For each incorrect key value, an error is injected into the circuit for only one input pattern, leading to an incorrect output for the specific pattern. Assuming that $F(I)$ is the original circuit, the output O of the circuit locked using SARLock can be presented as $O = F(I) \oplus ((I == K) \oplus (I == k_s))$, where K denotes the key inputs, and k_s is the correct key value (refer to Sect. 5.2 for details).

Anti-SAT The Anti-SAT block shown in Fig. 2.3b is constructed using two complementary blocks, $B_1 = g(X, K_{l1})$ and $B_2 = \overline{g(X, K_{l2})}$ [22]. These blocks share the same inputs X but are locked with different keys K_{l1} and K_{l2}. The outputs of B_1 and B_2 drive an AND gate to produce the output signal Y. The two blocks produce complementary outputs when the correct key value is applied; for all inputs, $Y = 0$, leading to a correct output. For an incorrect key value, the output of B_1 and B_2 is 1 for a specific input pattern; for that pattern, $Y = 1$, leading to an incorrect output. Assuming that Anti-SAT protects one of the primary outputs of the original circuit $F(I)$, the protected output O can be represented as $O = F(I, K_{l0}) \oplus (g(X \oplus K_{l1}) \wedge \overline{g(X \oplus K_{l2})})$, where K_{l0} represents the key for the logic locked circuit (refer to Sect. 5.3 for details).

AND-Tree Detection (ATD) As opposed to Anti-SAT that integrates external point functions with the original netlist, ATD identifies such structures inside an original netlist and reuses them to decrease the implementation cost [10]. Once an AND (OR)-tree has been identified in a netlist, it is locked by inserting XOR/XNOR key gates at its inputs (refer to Sect. 5.4 for details).

Tenacious and Traceless Logic Locking (TTLock) Both SARLock and Anti-SAT are vulnerable to structural attacks since they implement the original function as is [31]. In TTLock, the original logic cone $F(I)$ is modified for exactly one input pattern i_s to hide the true implementation from an attacker and thwart structural attacks [31]. The output of the logic cone for the *protected* input pattern is then restored using a protection circuit that is essentially a comparator block. TTLock is illustrated in Fig. 2.3c. Upon removal attack, the attacker retrieves a netlist which differs from the original netlist for exactly one input pattern. A limitation of TTLock is that it protects only one input pattern, resulting in (1) the minimal removal attack resilience, and (2) the minimal error injection. For any input pattern, only two key values inject an error into the circuit. (refer to Sect. 9.2 for details).

SFLL-HD Whereas TTLock modifies and thus protects only one input pattern, SFLL-HD allows to efficiently protect a large number of input patterns [30], leading to a higher removal attack resilience and a higher error injection rate. SFLL-HD is able to protect $\binom{k}{h}$ input patterns that are of Hamming distance h from the k-bit secret key. Only one k-bit secret key is stored in the tamper-proof memory. As depicted in Fig. 2.3c, a single comparator is used along with the Hamming distance checker. With increasing h, the number of protected patterns increases binomially. The SAT attack resilience decreases logarithmically with increasing number of protected patterns. For $h = 0$, SFLL-HD is equivalent to TTLock (refer to Sect. 9.3 for details).

SFLL-Flex SFLL-HD is suitable for general applications where it is useful to protect an arbitrary set of input patterns. However, in certain applications, a specified set of input patterns or a range of input patterns needs to be protected. SFLL-flex allows to compactly represent the patterns-to-be-protected using a small set of input cubes [30].[1] The input cubes are stored on an on-chip look-up table as illustrated

---

[1] Input cubes refer to partially-specified input patterns; some input bits are set to logic-0's or logic-1's, while other input bits are don't cares (x's).

in Fig. 2.3d. Only upon loading the correct cubes to the LUT, the circuit becomes functional (refer to Sect. 9.4 for details).

While point function-based techniques thwart the SAT attack by increasing the number of iterations required for the attack to succeed, ORF-Lock makes use of one-way functions to increase the time for individual attack iterations (refer to Sect. 8.3 for details). CycSAT is another SAT attack resilient technique that introduces cycles into a locked circuit, rendering the traditional SAT attack ineffective (refer to Sect. 8.1 for details).

2.4 Logic Locking Attacks

2.4.1 Algorithmic Attacks

Having already introduced the sensitization attack and the SAT attack in Sect. 2.1, we introduce only the CP attack and the CycSAT attack here.

Circuit Partitioning (CP) Attack The CP attack operates in a divide-and-conquer fashion [9]. The attack divides a circuit into logic cones and targets individual logic cones using brute-force. A logic cone is a sub-circuit consisting of gates that are in the transitive fan-in of a specific primary output. Depth-first search may be used to construct a logic cone. This divide-and-conquer approach is also utilized by the DPA attack [24].

CycSAT The CycSAT attack circumvents cyclic logic locking. The attack builds on top of the SAT attack; it adds additional constraints to the SAT formula that help eliminate the cycles from the netlist [32].

2.4.2 Structural Attacks

Signal Probability Skew (SPS) Attack The basic Anti-SAT block comprises an AND-tree and a NAND-tree, whose outputs are skewed towards 0 and 1, respectively [22]. The SPS attack exploits these structural traces—the skew in the signal probabilities—to locate and isolate the protection logic [26]. The attack becomes less effective in the presence of structural/functional obfuscation (refer to Sect. 7.1 for details).

AppSAT-Guided Removal (AGR) Attack To isolate the Anti-SAT block that has been obfuscated with additional XOR and multiplexer key gates, the AGR attack integrates AppSAT with simple post-processing steps [29] (refer to Sect. 7.2 for details).

Sensitization-Guided SAT (SGS) Attack The SGS attack exploits the security vulnerabilities of ATD to weaken the security it promises. The attack leverages the bias in the input distribution of ATD to reduce the computational effort (refer to Sect. 7.3 for details).

Bypass Attack The Bypass attack adds *bypass* circuitry around a locked netlist [23] to restore its correct functionality. The attack assigns a random key to the locked netlist and determines the DIP(s) for which the netlist outputs are incorrect; the bypass circuit is then designed to restore the output for the computed DIPs (refer to Sect. 7.4 for details).

2.4.3   Side-Channel Attacks

Differential Power Analysis (DPA) Attack The DPA attack is one of the most powerful side-channel attacks that have been used to break most of the cryptographic algorithms [7]. DPA attack on logic locking is launched by collecting power samples and the ciphertext output from the IC under attack for a large set of plaintext inputs [24]. The collected samples are then analyzed using statistical analysis yielding a differential trace, which tends to be high for the correct key value and zero for the incorrect key values [24]. The DPA attack may be launched on the individual logic cones in a divide-and-conquer approach (refer to Sect. 10.1 for details).

Test-Data Attacks While the regular attacks on logic locking rely on input/output observations recorded from a functional IC, test-data attacks, such as the hill climbing attack [12] and the test-data mining attack [28], extract secret information from the test data. Test data is generated at the design house using automatic test pattern generation (ATPG) tools, sent to the foundry/dedicated test facility, and utilized during manufacturing test to classify ICs as faulty/fault-free. The same ATPG tools can be used by an attacker in the test facility to leak the secret logic locking key (refer to Sects. 10.2 and 10.3 for details).

De-synthesis Attack The de-synthesis attack relies on the observation that the locked and the original netlist should be similar in terms of the type and count of gates [11]. The attack re-synthesizes the locked netlist with a random key and then uses hill climbing search to find the key value that yields the maximum similarity between the locked netlist and the re-synthesized netlist (refer to Sect. 10.4 for details).

2.5   Attack-Defense Matrix

Table 2.1 presents the resiliency of various logic locking countermeasures against each of the attacks mentioned above. It can be observed, for example, that the SAT attack [21] breaks all Pre-SAT logic locking defenses. Compound logic locking that

Table 2.1 Attack resiliency of logic locking techniques

Attack	RLL [17]	FLL [1, 15]	SLL [14]	AntiSAT [22]	SARLock [25]	SFLL [30]	Compound [22, 25]	Cyclic [18]	ORF-Lock [27]
Sensitization [14]	✗	✗	✓	✓	✓	✓	✓	✓	✓
SAT [21]	✗	✗	✗	✓	✓	✓	✓	✓	✓
CycSAT [32]	✗	✗	✗	✓	✓	✓	✓	✗	✓
AppSAT [19]	✗	✗	✗	✓	✓	✓	✗*	✓	✓
Double-DIP [20]	✗	✗	✗	✓	✓	✓	✗*	✓	✓
Removal/SPS [26]	✓	✓	✓	✗	✗	✓	✗	✓	✗
AGR [29]	✓	✓	✓	✗	✗	✓	✗	✓	✓
Bypass [23]	✓	✓	✓	✗	✗	✓	✓	✓	✓
Test-data mining [28]	✗	✗	✗	✓	✓	✓	✓	✓	✓
Hill climbing [12]	✗	✗	✓	✓	✓	✓	✓	✓	✓
DPA [24]	✗	✗	✗	✓	✓	✓	✓	✓	✓
Desynthesis [11]	✗	✗	✗	✓	✓	✓	✓	✓	✓

✗ denotes that a technique is vulnerable the attack, ✓ denotes resilience against the attack, and ✗* denotes partial attack success

integrates point function-based locking with Pre-SAT defenses can be circumvented using the approximate attacks such as AppSAT or Double-DIP. Among the SAT resilient logic locking techniques, only SFLL resists all known attacks on logic locking. Every other technique is vulnerable to at least one attack.

2.6   The Ever-Evolving Metrics

The metrics used to evaluate the effectiveness of logic locking have also evolved along with the logic locking techniques. Many of the metrics quantify the security of a given defense against specific attacks. Following is a summary of the metrics deployed over the years:

- **Output corruptibility (OC).** This metric quantifies protection against black-box usage of an IC; it was first used to highlight the advantage of FLL [15] over RLL [17]. An RLL netlist has the disadvantage that it may produce correct output for a large fraction of input/key combinations. Output corruptibility (OC) represents an estimate of the Hamming distance between the correct output and the incorrect output obtained on supplying random incorrect keys. The OC may be computed by applying P random input patterns, each with Q random key values, to the locked netlist L and observing its M-bit output O_L. OC represents the average percentage Hamming distance between the correct output O_F of the original function F and the output O_L obtained from the locked netlist.

  For a locked netlist L with

  $$OC = \frac{1}{P \times Q \times M} \sum_{i=1}^{P} \sum_{j=1}^{Q} HD(O_F(I_i), O_L(I_i, K_j)) \times 100\% \qquad (2.1)$$

  Here, I_i and K_j represent a random input pattern and a random key value, respectively. An OC value of 50% is ideal as it maximizes the ambiguity for an attacker [15].

- **Output error rate (OER).** Related to OC is the concept of OER. While OC represents the average percentage Hamming distance between the reference and the observed output, OER represents the percentage of input patterns that lead to an incorrect output. OER only considers if any of the M output bits is incorrect, whereas OC also takes into the number of incorrect output bits.

$$OER = \frac{1}{P} \sum_{i=1}^{P} a_i \times 100\% \qquad (2.2)$$

Where,

$$a_i = \begin{cases} 1, \text{if} \quad (\sum_{j=1}^{Q} HD(O_F(I_i), O_L(I_i, K_j))) \geq 1 \\ 0, \text{otherwise} \end{cases} \qquad (2.3)$$

OER is often used to quantify the error injected by various point function-based techniques. The terms OC and OER have been used interchangeably in the literature.

- **Clique size**. Clique size quantifies the interference among the key gates and represents the resilience against the sensitization attack; the metric was introduced in the context of SLL [14]. Key gates that protect one another such that an individual key bit cannot be sensitized to primary outputs without knowing/controlling the values of remaining key bits form a clique. The number of key gates that form the largest clique in a circuit is referred to as clique size [27].
- **Number of distinguishing input patterns (#DIPs)**. Distinguishing input patterns are the special input patterns that are computed by the SAT attack and help refine the key search space [21]. In each iteration, the attack applies computes a DIP, which along with the corresponding output of the functional IC, helps distinguish the feasible key values from the infeasible ones. The computational effort of the attack can be represented in terms of the number of distinguishing input patterns required for a successful SAT attack.
- **Percentage of key bits recovered**. Certain attacks may succeed only partially and extract the values for only a subset of the key bits [12, 21, 24]. The success rate of such attacks is quantified in terms of the percentage of the key bits retrieved correctly. A powerful attack will recover a significant percentage of the key bits, enabling the attacker to brute-force on the remaining key bits.
- **Execution time**. The execution time of an attack can also be used to demonstrate the resilience of a logic locking technique against a specific attack [21]. Conventionally, the execution time is empirically recorded for smaller key sizes and then extrapolated to show the trend for larger key sizes [22, 25, 30].

To conclude, this chapter presented a summary of logic locking attacks, defenses, and the metrics used to evaluate logic locking algorithms. The chapter emphasized the relationship between the logic locking attacks and defenses in the form of an attack-defense matrix. It also presented a timeline of major developments in combinational logic locking. Overall, the chapter serves as a preview into the contents of the forthcoming chapters of the book.

References

1. Baumgarten A, Tyagi A, Zambreno J (2010) Preventing IC piracy using reconfigurable logic barriers. IEEE Des Test Comput 27(1):66–75
2. Chakraborty RS, Bhunia S (2009) HARPOON: an obfuscation-based SoC design methodology for hardware protection. IEEE Trans Comput Aided Des Integr Circuits Syst 28(10):1493–1502
3. Colombier B, Bossuet L, Hély D (2017) Centrality indicators for efficient and scalable logic masking. In: EEE computer society annual symposium on VLSI, pp 98–103
4. Dupuis S, Ba P, Natale GD, Flottes M, Rouzeyre B (2014) A novel hardware logic encryption technique for thwarting illegal overproduction and hardware trojans. In: IEEE international on-line testing symposium, pp 49–54

5. Karousos N, Pexaras K, Karybali IG, Kalligeros E (2017) Weighted logic locking: a new approach for IC piracy protection. In: IEEE international symposium on on-line testing and robust system design, pp 221–226
6. Khaleghi S, Zhao KD, Rao W (2015) IC piracy prevention via design withholding and entanglement. In: Asia Pacific design automation conference, pp 821–826
7. Kocher P, Jaffe J, Jun B (1999) Differential power analysis. In: Advances in cryptology. Springer, Berlin, pp 388–397
8. Koushanfar F (2012) Provably secure active IC metering techniques for piracy avoidance and digital rights management. IEEE Trans Inf Forensics Secur 7(1):51–63
9. Lee YW, Touba N (2015) Improving logic obfuscation via logic cone analysis. In: Latin-American test symposium, pp 1–6
10. Li M, Shamsi K, Meade T, Zhao Z, Yu B, Jin Y, Pan D (2016) Provably secure camouflaging strategy for IC protection. In: IEEE/ACM international conference on computer-aided design, pp 28:1–28:8
11. Massad M, Zhang J, Garg S, Tripunitara M (2017) Logic locking for secure outsourced chip fabrication: a new attack and provably secure defense mechanism. CoRR abs/1703.10187, http://arxiv.org/abs/1703.10187
12. Plaza S, Markov I (2015) Solving the third-shift problem in IC piracy with test-aware logic locking. IEEE Trans Comput Aided Des Integr Circuits Syst 34(6):961–971
13. Rajendran J, Pino Y, Sinanoglu O, Karri R (2012) Logic encryption: a fault analysis perspective. In: Design, automation and test in Europe, pp 953–958
14. Rajendran J, Pino Y, Sinanoglu O, Karri R (2012) Security analysis of logic obfuscation. In: IEEE/ACM design automation conference, pp 83–89
15. Rajendran J, Zhang H, Zhang C, Rose G, Pino Y, Sinanoglu O, Karri R (2015) Fault analysis-based logic encryption. IEEE Trans Comput 64(2):410–424
16. Roy JA, Koushanfar F, Igor L (2008) EPIC: ending piracy of integrated circuits. In: Design, automation, and test in Europe, pp 1069–1074
17. Roy J, Koushanfar F, Markov IL (2010) Ending piracy of integrated circuits. IEEE Comput 43(10):30–38
18. Shamsi K, Li M, Meade T, Zhao Z, Pan DZ, Jin Y (2017) Cyclic obfuscation for creating sat-unresolvable circuits. In: ACM Great Lakes symposium on VLSI, pp 173–178
19. Shamsi K, Li M, Meade T, Zhao Z, Z D, Jin Y (2017) AppSAT: approximately deobfuscating integrated circuits. In: IEEE international symposium on hardware oriented security and trust, pp 95–100
20. Shen Y, Zhou H (2017) Double DIP: Re-Evaluating Security of Logic Encryption Algorithms. Cryptology ePrint Archive, Report 2017/290, http://eprint.iacr.org/2017/290
21. Subramanyan P, Ray S, Malik S (2015) Evaluating the security of logic encryption algorithms. In: IEEE international symposium on hardware oriented security and trust, pp 137–143
22. Xie Y, Srivastava A (2016) Mitigating SAT attack on logic locking. In: International conference on cryptographic hardware and embedded systems, pp 127–146
23. Xu X, Shakya B, Tehranipoor M, Forte D (2017) Novel bypass attack and BDD-based tradeoff analysis against all known logic locking attacks. In: International conference on cryptographic hardware and embedded systems, pp 189–210
24. Yasin M, Mazumdar B, Ali SS, Sinanoglu O (2015) Security analysis of logic encryption against the most effective side-channel attack: DPA. In: IEEE international symposium on defect and fault tolerance in VLSI and nanotechnology systems, pp 97–102
25. Yasin M, Mazumdar B, Rajendran J, Sinanoglu O (2016) SARLock: SAT attack resistant logic locking. In: IEEE international symposium on hardware oriented security and trust, pp 236–241
26. Yasin M, Mazumdar B, Sinanoglu O, Rajendran J (2016) Security analysis of anti-SAT. In: IEEE Asia and South Pacific design automation conference, pp 342–347
27. Yasin M, Rajendran J, Sinanoglu O, Karri R (2016) On improving the security of logic locking. IEEE Trans Comput Aided Des Integr Circuits Syst 35(9):1411–1424

28. Yasin M, Saeed SM, Rajendran J, Sinanoglu O (2016) Activation of logic encrypted chips: pre-test or post-test? In: Design, automation test in Europe, pp 139–144
29. Yasin M, Mazumdar B, Sinanoglu O, Rajendran J (2017) Removal attacks on logic locking and camouflaging techniques. IEEE Trans Emerg Top Comput. https://doi.org/10.1109/TETC.2017.2740364
30. Yasin M, Sengupta A, Nabeel MT, Ashraf M, Rajendran J, Sinanoglu O (2017) Provably-secure logic locking: from theory to practice. In: ACM/SIGSAC conference on computer & communications security, pp 1601–1618
31. Yasin M, Sengupta A, Schafer B, Makris Y, Sinanoglu O, Rajendran J (2017) What to lock?: functional and parametric locking. In: Great Lakes symposium on VLSI, pp 351–356
32. Zhou H, Jiang R, Kong S (2017) CycSAT: SAT-based attack on cyclic logic encryptions. In: IEEE/ACM international conference on computer-aided design, pp 49–56

Chapter 3
Pre-SAT Logic Locking

Abstract This chapter focuses on the Pre-SAT logic locking, presenting three techniques, RLL, FLL, and SLL, in addition to describing the sensitization attack. RLL is the earliest known logic locking technique that was introduced to thwart IC piracy. FLL improves upon RLL and prevents the black-box usage of an IC. However, both RLL and FLL remain susceptible to the sensitization attack, which retrieves the key bits from a functional IC in a divide-and-conquer fashion. SLL thwarts the sensitization attack by inserting key gates that exhibit strong interference among themselves and make it hard to retrieve key bits on an individual basis.

This chapter is about the Pre-SAT logic locking techniques. Section 3.1 introduces RLL, the first-ever logic locking technique. Section 3.2 describes FLL, which improves upon RLL to prevent the black-box usage of an IC. Section 3.3 focuses on the sensitization attack, the first algorithmic attack on logic locking. Section 3.3 presents SLL as a countermeasure to the sensitization attack.

3.1 Random Logic Locking

3.1.1 Motivation

RLL, also referred to as EPIC, was introduced by Roy et al. [8] in 2008. The motivation behind RLL is to hide (lock) the true functionality of a Boolean circuit from an attacker who intends to steal the valuable IP. EPIC aims at securing the IP from untrusted agents present anywhere in the IC design flow except for the trusted design house. The IP owner locks the original netlist by inserting XOR/XNOR key gates at random locations in the netlist. The locked netlist that passes through the untrusted design and fabrication phases is considered non-functional as it will produce incorrect outputs without the application of the correct key. The IP owner, who has the correct key, can activate a fabricated IC by loading the correct key to the

M. Yasin et al., *Trustworthy Hardware Design: Combinational Logic Locking Techniques*, Analog Circuits and Signal Processing, https://doi.org/10.1007/978-3-030-15334-2_3

on-chip memory in a trusted facility. EPIC also allows remote activation of ICs at an untrusted foundry by enabling secure communication with a fabricated IC through public-key cryptography [9].

3.1.2 The RLL Algorithm

The RLL algorithm, illustrated in Algorithm 1, has two phases. In the first phase, k XOR/XNOR key gates are inserted into a netlist, where k denotes the key size. In the second phase, the types of key gates are adjusted to either XNOR or XOR to break the one-to-one mapping between the key gate type and the secret key value.

Phase 1 The RLL algorithm operates on a gate level netlist, comprising Boolean gates. The first phase inserts XOR/XNOR key gates at the output of k randomly selected gate locations. If the ith key bit k_i is 0, an XOR key gate is inserted; otherwise, an XNOR key gate is inserted.

Example Figure 3.1b shows a locked netlist with three XOR/XNOR key gates; the correct key value is 110. Note that in the example, there is a one-to-one mapping between the key gate type and the key value.

Algorithm 1: RLL algorithm [8, 9]

Input   : Original netlist N, Secret key K
Output: Locked netlist N_{lock}

1 $g \leftarrow$ #gates in N
2 $N_{lock} \leftarrow N$
   // **Phase 1: Key Gate Location Selection**
3 **for** $i = 1$ *to* k **do**
4       $r = \text{rand}(1,g)$ // Select a random gate location
5   insert_keygate(N_{lock}, r, k_i) // Insert an XOR/XNOR key gate at the output of rth gate
6 **end**
   // **Phase 2: Modification**
7 **for** $i = 1$ *to* k **do**
8       $inv = \text{rand}(0,1)$ // Flip a coin
9   **if** $inv == 1$ **then**
10   insert_inverter(N_{lock}, kg_i) // Insert an inverter at the output of the ith key gate
11   change_key_gate_type (kg_i) // adjust the type of the ith key gate, convert an XNOR to an XOR and vice versa
12   merge_and_move_inverters (N_{lock})
13   **end**
14 **end**

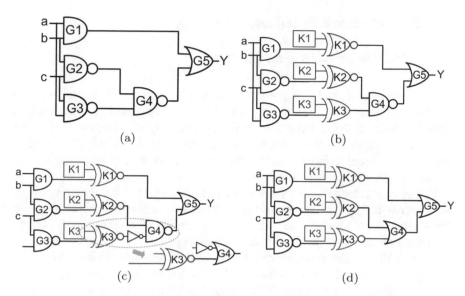

Fig. 3.1 Locking a netlist using RLL. (**a**) The original netlist. (**b**) Phase 1: inserting XOR/XNOR gates into the netlist. The correct key for the locked netlist is 110. (**c**) Phase 2: Inserting an inverter at the output of key gate K3 and changing the gate type to XNOR. (**d**) The final netlist obtained after bubble-pushing the inverters

Phase 2 The one-to-one mapping between the key gate type and the key bit value must be broken to make the inference of key bit values computationally harder for an attacker. The second phase of the RLL algorithm accomplishes this objective by (1) inserting inverters at the output of randomly selected key gates and (2) changing the type of the affected key gates, i.e., converting XORs to XNORs and vice versa. If an inverter is placed next to an already existing inverter, effectively nullifying the previous inversion, both inverters may be replaced with a wire. Alternatively, the inverters may be bubble-pushed using De-Morgan's laws; this process may change the Boolean function of certain gates.

With these modifications, an attacker cannot figure out whether a particular inverter is part of the original circuit or it has been introduced during the modification phase. Furthermore, he/she cannot determine if a certain gate was present in the original netlist as is or it has been modified.

Example Figure 3.1c shows an intermediate netlist obtained by inserting inverters after selected gates. Only the key gate K3 is modified during this example ($inv = 1$ only for $i = 3$). By applying De-Morgan's laws, the newly introduced inverter is absorbed into the gate G4, transforming it from a NAND gate to an OR gate. During this process, an additional inverter is introduced at the first input of G4; this inverter is incorporated into key gate K2, turning it into an XOR gate as illustrated in Fig. 3.1d. While the types of the three key gates are now XNOR, XOR and XNOR, the correct key for the circuit is still 110.

3.2   Fault-Analysis Based Logic Locking

3.2.1   Motivation: Black-Box Usage

A drawback of RLL is that it does not take into account the impact of key gate
location on the outputs. Consequently, an RLL circuit may produce correct output
even for incorrect key values. The output corruptibility of RLL turns out to be
low, which enables this black-box usage of a locked IC. As already illustrated in
Table 1.2, for specific input patterns, the locked netlist may produce the correct
output regardless of the key value.

   FLL aims at overcoming the shortcomings of RLL, controllably achieving a high
output corruptibility. FLL inserts key gates at the most influential gate locations in
the netlist, i.e., the locations that exhibit the highest impact on the primary outputs
and render the circuit output maximally incorrect upon the application of incorrect
key values. FLL makes extensive use of the VLSI test principles of fault excitation,
propagation, and masking (see Appendix A for a detailed description). By relating
these principles and the impact of inserting XOR/XNOR key gates in a netlist,
FLL develops the notion of fault impact metric, which helps select the locations
for inserting key gates.

3.2.2   Logic Locking and Fault Analysis

Table 3.1 depicts how fundamental VLSI testing principles are related to logic
locking. Figure 3.2 further elaborates on these relations.

Fault Excitation   The impact of applying an incorrect key is analogous to excitation
of a stuck-at fault at the key input. Applying an incorrect key ($K1 = 1$) in Fig. 3.2b,
results in negating A, which is the same as exciting a s-a-0 fault (for $A = 1$) or a
s-a-1 fault (for $A = 0$).

Fault Propagation   Upon excitation, a fault may be propagated to a primary output,
generating an incorrect output. Similarly, applying incorrect key values may lead to
an incorrect output for certain input patterns. An incorrect output is observed when
non-controlling values are applied on the side-inputs of the gates in the propagation
path of the fault (a key gate supplied with an incorrect key).

Table 3.1 Relating the
principles of VLSI testing to
logic locking [6]

VLSI testing	Logic locking
Fault excitation	Invalid key
Fault propagation	Invalid output
Multiple faults	Multiple invalid keys
Fault masking	Self-cancellation of invalid keys

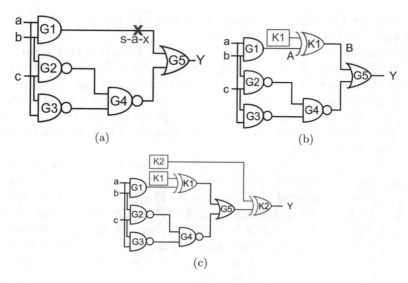

Fig. 3.2 Relation between VLSI testing and logic locking. (**a**) Fault excitation ⇔ Incorrect Key, (**b**) Fault propagation ⇔ Incorrect output, and (**c**) Fault masking ⇔ self-cancellation of incorrect keys

Fault Masking The effect of applying multiple incorrect keys is the same as exciting multiple stuck-at faults. However, the excited faults may cancel/mask one another, similar to multiple incorrect keys nullifying each other's impact. When an incorrect key value 11 is applied to the circuit in Fig. 3.2c, the error introduced at the output of key gate K1 may be cancelled by the error at key gate K2.

3.2.3   The FLL Algorithm

The discussion above hints that for maximum output corruptibility, key gates must be inserted at the locations that affect the most of the outputs for most of the input patterns upon excitation of faults (application of incorrect key values). This influence of a gate G can be represented using the fault impact FI_G.

$$FIG = (P_0 \times O_0) + (P_1 \times O_1) \tag{3.1}$$

Where P_x denotes the number of patterns that detect a s-a-x fault at the output of the gate G, and O_x denotes the "cumulative" number of output bits that are affected by that s-a-x fault. Recall that in Eq. 2.1, OC depends on two factors, (1) the number of patterns that induce a fault at the outputs (think in terms of the OER), and (2) the number of output bits that exhibit an error (think in terms of the HD). O_x is similar to the HD; it represents the "cumulative" number of output bits that are affected when there is a s-a-x fault at G.

To compute fault impact, FLL selects $P = 1000$ random input patterns and conducts *fault simulation* on the target netlist. Traditionally fault simulation has been used to measure the test quality, i.e., what percentage of total faults in a circuit can be detected by a given set of test vectors (input patterns). In the context of FLL, fault simulation is targeted to individual gates; the objective is to determine the "quality" of a given gate location and assess its contribution towards improving the overall OC of the circuit.

The objective of FLL is to build a locked circuit for which the application of incorrect keys will lead to an OC of 50%, which leads to the maximum ambiguity for an attacker. As depicted in Algorithm 2, the FLL algorithm has two phases. In the key gate location selection phase, k XOR/XNOR key gates are inserted at the locations that exhibit the highest fault impact. Upon the insertion of the ith key gate, the netlist is updated and the fault impact is recomputed. In the modification phase, the types of selected key gates are changed to break the one-to-one mapping between the key gate type and the key value.

Example Let us consider the application of the first iteration of FLL to the c17 ISCAS benchmark circuit shown in Fig. 3.3. Table 3.2 lists P_0, O_0, P_1, O_1, and FI_G

Algorithm 2: FLL algorithm [4, 6]

Input   : Original netlist N, Secret key K
Output: Locked netlist N_{lock}

1 $N_{lock} \leftarrow N$
2 $P \leftarrow 1000$
    // Phase 1: Key Gate Location Selection
3 **for** $i = 1$ *to* k **do**
4   apply_test_patterns(P)
5   **foreach** *gate* $g \in N_{lock}$ **do**
6 | compute_FaultImpact(g)
7   **end**
8   Select the gate g_{maxfi} with the highest FaultImpact
9   insert_keygate(N_{lock}, g_{maxfi}, k_i)
10 **end**
    // Phase 2: Modification
11 **for** $i = 1$ *to* k **do**
12   inv = rand(0,1) // Flip a coin
13   **if** inv == 1 **then**
14   insert_inverter(N_{lock}, kg_i) // Insert an inverter at the output of the ith key gate
15   change_key_gate_type(kg_i) // adjust the type of the ith key gate, convert an XNOR to an XOR and vice versa
16   merge_and_move_inverters (N_{lock})
17   **end**
18 **end**

Fig. 3.3 c17 ISCAS benchmark circuit

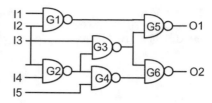

Table 3.2 Fault impact for different gates in c17 ISCAS benchmark circuit

Gate (G)	P_0	O_0	P_1	O_1	FI_G
G1	444	958	200	1030	631,352
G2	542	1062	194	962	762,232
G3	617	902	329	1043	899,681
G4	458	938	186	1006	616,720
G5	556	1070	444	958	1,020,272
G6	542	1022	458	938	983,528

for all the gates in the circuit; these values are calculated by simulating $P = 1000$ random input patterns. The most influential location of inserting the first key gate turns out to be the gate $G5$; the corresponding OC is 23%. With the insertion of more key gates in subsequent iterations, the OC improves.

Figure 3.4 illustrates the advantage of FLL over RLL by plotting output corruptibility OC as a function of the key size k. Most of the RLL circuits achieve a low OC (<40%) even upon insertion of 64 key gates. Most FLL circuits, however, approach the target 50% OC value even with a key size of 20.

3.3 Sensitization Attack

As mentioned in Sect. 2.1, the sensitization attack is the first attack against any logic locking technique [5]. The attack makes use of the VLSI testing principle of sensitization to recover the secret key for RLL and FLL circuits. Sensitization of a net w to an output O implies that the value of w is observable on the output O either as is or in the negated form. More specifically, there exists a bijective mapping between the values of the net and the values observed at a primary output.

3.3.1 Threat Model

The attack operates under the following assumptions:

- The designer is trusted, i.e., the personnel and the tools used in the design house are trustworthy.
- The foundry and the end-user are untrusted.

Fig. 3.4 Output
corruptibility vs. key size for
(**a**) RLL [9] and (**b**) FLL [6]

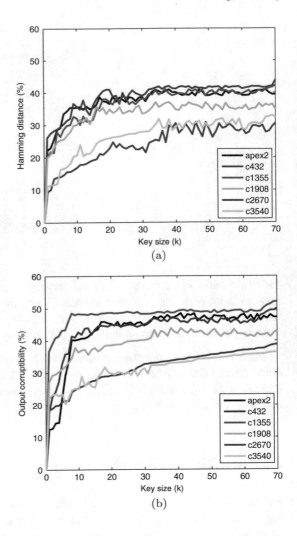

- The attacker has access to (1) a reverse-engineered netlist and (2) a functional chip that embeds the secret key.
- The attacker knows the logic locking algorithm, as well as the location of the protection circuitry. The only unknown is the secret key.

This threat model has been widely adopted in the logic locking literature and can be considered as the standard threat model for logic locking [7, 11].

3.3.2 Attack Algorithm

The sensitization attack exploits the fact that there is minimal interference among the key gates in RLL and FLL. In the extreme case, all key gates may be isolated

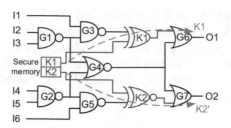

Fig. 3.5 An illustration of the sensitization attack. By setting the key bit K2 to x, an attacker computes the pattern 100xxx. By applying this pattern to the functional IC, the value of key bit K2 can be observed on the primary output O1. In this particular example, it is also possible to sensitize both K1 and K2 to outputs O1 and O2, respectively, by applying the input pattern 100001. K1 propagates as is, while K2 is negated

from one another, making it possible to retrieve the key bit values on an individual basis. The attack is an iterative algorithm; the following steps are repeated in each iteration of the attack:

1. The attacker analyzes the locked netlist and computes an attack pattern that can sensitize one (or more) of the key bits to a primary output. Existing ATPG tools can be used for computing the attack patterns. When targeting a specific key bit, e.g., K1 in Fig. 3.5, an attacker can set the other key bits to unknown x's and conduct ATPG for detecting a stuck-at fault at the key input K1.
2. The computed attack pattern is applied to the functional IC and its output is recorded. By analyzing the observed IC response, an attacker can determine the value of the targeted key bit.

Example Consider the key bit K1 in Fig. 3.5, which is targeted during the first iteration of the attack. The key bit will be sensitized to output O1 only if the value at the other input of gate G6 is 0. G4 can be made 0 by setting $I1 = 1$, $I2 = 0$, and $I3 = 0$. Note that ATPG tools such as Atalanta [3] or Synopsys TetraMax can be used to generate such attack patterns. The attacker provides the locked netlist and a set of constraints to the tool. In this example, the value of the other key bit K2 is unknown and it can be denoted as an x. By setting K2 to x, the attacker finds a pattern which can detect a stuck-at fault (either s-a-0 or s-a-1) at the key input K1. The tool generates the pattern 100xxx. The attacker applies this pattern to the functional IC and observes O1, which must be the same as K1. If $O1 = 1$, then $K1 = 1$; otherwise, $K1 = 0$. If the tool generates the pattern 000xxx, the output of the NAND gate G3 (which is the other input of the key gate) is 1 and the key bit K1 is sensitized to O1 in negated form; if $O1 = 1$, then $K1 = 0$ and vice versa. In the second iteration, the attack targets key bit K2. The attack pattern xxx001 sensitizes K2 to O2 in the negated form; if $O2 = 1$, then $K2 = 0$ and vice versa. Overall, the sensitization attack retrieves all the key bits correctly.

Attacking Other Interference Configurations The previous example considers only the isolated key gates with no interference among them, enabling an attacker to target the key gates on an individual basis. However, the attack can be adapted to cater to other key gate configurations (refer to [5] and [12] for details). The sensitization attack can be thwarted using SLL [5] as discussed in the next section.

3.4   Strong Logic Locking

3.4.1   Basic Idea

The discussion in the previous section indicates that RLL and FLL remain vulnerable to the sensitization attack [5]. Strong logic locking thwarts the sensitization attack by inserting key gates in a way that maximizes the interference among the key gates and prevents sensitization of the key bits on an individual basis. The basic premise is that with increased interference among the key gates, an attacker is forced to brute-force an exponentially increasing number of key combinations [5]. SLL inserts *pairwise secure* key gates that protect one another in a netlist.

3.4.2   Pairwise Security

In this section, we elaborate on the concept of pairwise secure key gates. Consider the netlist in Fig. 3.6, which has two key gates, K1 and K2. Both key gates interfere in each other's path to the primary outputs. More importantly, it is not possible for an attacker to sensitize either K1 or K2 to a primary output on an individual basis. Let's assume that the attacker wants to sensitize the key bit K2 to one of the primary outputs. To propagate the value of K2 through the NOR gate G4, the other input of G4 must be 0; otherwise, the output of G4 will be 1, effectively masking K2. However, without knowing the value of key bit K1, one cannot control the value of net A. Thus, K1 protects K2 by hampering its sensitization. Following the same

Fig. 3.6 An illustration of pairwise secure key gates in SLL [12]. None of the key bits K1 or K2 can be sensitized to a primary output unless the value of the other keybit is known or its effect can be muted

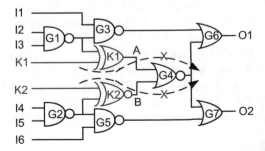

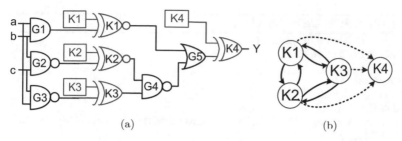

Fig. 3.7 (**a**) A locked circuit with four key gates. (**b**) Interference graph of the key gates

argument, we observe that K2 protects K1. Overall, K1 and K2 protect each other and are referred to as *pairwise secure* key gates.

Notation Let us assume $P_{K2|K1=x}$ denotes the set of primary input patterns (P) that can sensitize key bit $K2$ to a primary output when the value of key bit $K1$ is unknown. Then, $P_{K2|K1=x} = \emptyset$ implies that there exist no input patterns that can sensitize the $K2$ to a primary output if $K1 = x$. Similarly, $P_{K1|K2=x}$ denotes the set of input patterns that can sensitize $K1$ to a primary output given $K2 = x$. $K1$ and $K2$ are pairwise secure iff $P_{K2|K1=x} = \emptyset$ and $P_{K1|K2=x} = \emptyset$, i.e., an attacker cannot find an input pattern that can sensitize one of the key bits to a primary output without knowing or controlling the value of the other key bit. This way, he/she is forced to employ brute force on the values of the key bits.

Interference Graph Figure 3.7 demonstrates the concept of pairwise security using a directed graph. Each vertex in the graph in Fig. 3.7a represents a key gate and each edge specifies the relationship between two key gates. A solid edge from gate K1 to K2 denotes that K1 protects K2, i.e., $P_{K2|K1=x} = \emptyset$. A dashed edge represents that K1 does not protect K2, i.e., $P_{K2|K1=x} \neq \emptyset$. It can be observed, for example, that K1 and K2 are pairwise secure as indicated by solid edges between the two key gates. K1 and K4, however, are not pairwise secure since $P_{K4|K1=x} \neq \emptyset$. Even if $K1 = x$, its effect can be masked by setting $G4 = 1$, sensitizing K4 to Y.

Clique Size The notion of pairwise security is at the heart of SLL and it aims at maximizing the number of pairwise secure key gates. The metric used to quantify the security of SLL is clique size. A clique in a graph is a subgraph of nodes which are directly connected to one another, for example, a set of individuals who are all friends with one another. In the context of secure logic locking, a designer would want to insert key gates that are pairwise secure wrt one another; such a set of key gates forms a clique. All the gates in a clique are connected via solid edges. For the graph in Fig. 3.7b, the maximum clique size is three with the clique including the key gates K1, K2, and K3. K4 cannot be part of the clique since it is not pairwise secure with any of the other key gates.

Algorithm 3: SLL algorithm [12]

Input   : Original netlist, Secret key K
Output: Locked netlist

1 $SET_{KG} = \{\}$
2 cliqueCount = 0
3 $CLIQ_{cur}$ = Initialize()
4 **while** $|SET_{KG}| < k$ **do**
5   Find the set C of new candidates that can be pairwise secure with all key gates in $CLIQ_{cur}$
6   **For each** $Gate_i \in C$ **do**
7   **For each** $KG_j \in CLIQ_{cur}$ **do**
8   **if** $Gate_i$ *and* KG_j *are pairwise secure* **then**
9 | mark $Gate_i$ and KG_j as pairwise secure
10   **end**
11   **else**
12 | break // KG_j cannot be included in the clique
13   **end**
14   **end**
15   **if** $Gate_i$ *is marked pairwise secure with all key gates in* $CLIQ_{cur}$ **then**
16   Insert a key gate at the output of $Gate_i$
17   Add ($Gate_i$,cliqueCount) to SET_{KG}
18   Add $Gate_i$ to $CLIQ_{cur}$
19   KeyGateCount += 1
20   break // set of candidates needs to be re-computed
21   **end**
22   **else**
23   Mark $Gate_i$ as insecure
24   Mark $Gate_i$ as processed
25   **end**
26   **end**
27   **if** *all gates* $\in C$ *are processed* **then**
28     $CLIQ_{cur}$ = Initialize()
29   **if** $CLIQ_{cur} == \{\}$ **then**
30 | break // Terminate the algorithm
31   **end**
32   **end**
33 **end**
34

---

35 **KG$_{new}$ = function Initialize()**
36   Insert a key gate RG at a random unprocessed gate location
37   $CLIQ_{cur} = \{\}$
38   Add RG to $CLIQ_{cur}$
39   cliqueCount += 1
40   Add (RG,cliqueCount) to SET_{KG}
41   KeyGateCount += 1
42   Return $CLIQ_{cur}$

3.4.3   The SLL Algorithm

The SLL algorithm aims at maximizing the resilience against the sensitization attack by introducing key gate cliques in a circuit. Similar to the RLL and FLL algorithms, the SLL algorithm has two phases: a gate selection phase and a modification phase. Algorithm 3 presents only the gate selection phase; the modification phase has already been presented in Algorithm 1. As described in the *Initialize()* function in Algorithm 3, SLL inserts the first key gate at a random location and adds the key gate to the first clique $CLIQ_{cur}$. The algorithm then uses a heuristic approach to find the set C of candidate gate locations that are potentially pairwise secure with each of the key gates in the current clique. Only the gates in transitive fan-in and transitive fan-out of the current clique $CLIQ_{cur}$ are considered as candidates. Pairwise security checks are executed (using an ATPG tool) between a candidate gate $Gate_i$ and all the key gates in $CLIQ_{cur}$. If the gate $Gate_i$ is pairwise secure with all the gates in $CLIQ_{cur}$, it is added to the clique. If all the candidate gates have been marked as processed and the required key size is not achieved, the *Initialize()* function is invoked to create a new clique. The algorithm terminates when the required key size k is achieved or there are no more candidate gates to process.

3.5   The Variants of Basic Techniques

The chapter presented three basic logic locking techniques, RLL, FLL, and SLL. All three techniques insert key gates at selected locations in a netlist. The differences lie in the gate selection algorithm. However, researchers have also proposed logic locking techniques that can be considered as variants of the basic techniques. For example, Colombier et al. [1] make use of graph centrality indicators to find suitable key gate locations. Their approach tries to achieve the same OC as that achieved by FLL but in a scalable manner. FLL algorithm is iterative and runs extensive fault simulation in each iteration. Centrality indicator-based logic locking computes all the gate locations over the same graph in one run and avoids the iterative steps. Weighted logic locking [2] uses key gates with multiple key inputs to thwart the sensitization attack and simultaneously improve OC.

In summary, this chapter focused on the three basic logic locking techniques, RLL, FLL, and SLL. RLL inserts key gates at random locations in a netlist, FLL selects key gates locations using the notion of fault impact, and SLL inserts pairwise secure key gates that help thwart the sensitization attack. FLL and SLL algorithm rely highly on the principles of VLSI testing. These basic techniques, as well as their variants, remain vulnerable to the SAT attack [10], which is the focus of the next chapter.

References

1. Colombier B, Bossuet L, Hely D (2017) Centrality indicators for efficient and scalable logic masking. In: IEEE Computer Society annual symposium on VLSI, pp 98–103
2. Karousos N, Pexaras K, Karybali IG, Kalligeros E (2017) Weighted logic locking: a new approach for IC piracy protection. In: IEEE international symposium on on-line testing and robust system design, pp 221–226
3. Lee H, Ha D (1993) Atalanta: an efficient ATPG for combinational circuits. Technical Report
4. Rajendran J, Pino Y, Sinanoglu O, Karri R (2012) Logic encryption: a fault analysis perspective. In: Design, automation and test in Europe, pp 953–958
5. Rajendran J, Pino Y, Sinanoglu O, Karri R (2012) Security analysis of logic obfuscation. In: IEEE/ACM design automation conference, pp 83–89
6. Rajendran J, Zhang H, Zhang C, Rose G, Pino Y, Sinanoglu O, Karri R (2015) Fault analysis-based logic encryption. IEEE Trans Comput 64(2):410–424
7. Rostami M, Koushanfar F, Karri R (2014) A primer on hardware security: models, methods, and metrics. IEEE 102(8):1283–1295
8. Roy JA, Koushanfar F, Igor L (2008) EPIC: ending piracy of integrated circuits. In: Design, automation, and test in Europe, pp 1069–1074
9. Roy J, Koushanfar F, Markov IL (2010) Ending piracy of integrated circuits. IEEE Comput 43(10):30–38
10. Subramanyan P, Ray S, Malik S (2015) Evaluating the security of logic encryption algorithms. In: IEEE international symposium on hardware oriented security and trust, pp 137–143
11. Yasin M, Sinanoglu O (2017) Evolution of logic locking. In: IFIP/IEEE international conference on very large scale integration, pp 1–6
12. Yasin M, Rajendran J, Sinanoglu O, Karri R (2016) On improving the security of logic locking. IEEE Trans Comput Aided Des Integr Circuits Syst 35(9):1411–1424

Chapter 4
The SAT Attack

Abstract This chapter elaborates on the SAT attack, which breaks all pre-SAT logic locking techniques. The SAT attack is an oracle-guided attack that utilizes a SAT solver to compute attack patterns that refine the key search space iteratively. The SAT attack has changed the direction of logic locking research; developing efficient countermeasures against the attack is still an active area of research.

This chapter presents the SAT attack, which breaks all pre-SAT logic locking techniques. The SAT attack forms the most potent variant of key recovery attacks mounted to break basic combinational logic locking techniques [10]. The attack and its variants apply to both logic locking [10] and camouflaging [4, 5]. The attack uses the notion of Boolean satisfiability. Section 4.1 reviews the fundamental concepts of Boolean satisfiability. Section 4.2 presents the SAT attack algorithm. Section 4.3 elaborates on the effectiveness of the SAT attack against the pre-SAT logic locking techniques. Section 4.4 discusses the potential approaches to thwart the SAT attack. Section 4.5 presents a formal security analysis framework to quantify security of logic locking techniques against different classes of attacks including the SAT attack.

4.1 Preliminaries

4.1.1 Boolean Satisfiability

Boolean satisfiability (SAT) is the well-known problem of determining whether a Boolean formula is satisfiable, i.e. it evaluates to a logic 1. A value that makes the formula satisfiable is referred to as a satisfying assignment. For example, one of the satisfying assignment for the formula (a & b + !c) is (a,b,c) = (1,1,0). Typical SAT solvers accept a Boolean formula in the conjunctive normal form (CNF). A CNF formula is a conjunction (AND) of clauses,

© Springer Nature Switzerland AG 2020
M. Yasin et al., *Trustworthy Hardware Design: Combinational Logic Locking Techniques*, Analog Circuits and Signal Processing,
https://doi.org/10.1007/978-3-030-15334-2_4

where a clause is a disjunction (OR) of literals. A literal is either a variable (e.g. a), or the negation of a variable (e.g. !a). For example, the formula ((!a + b + !c) & (a + !b + !c) & c) contains three clauses; the & symbol may be omitted without any loss of generality. Each clause in a CNF formula must be satisfied for the formula to be satisfiable.

The research over the last two decades have resulted in powerful SAT solvers have been applied for solving formal verification, automatic test pattern generation, and many other challenging problems [1, 9]. For applying SAT solvers on combinational circuits, a circuit is first transformed to a CNF formula using the Tseitin transformation.

4.1.2 Tseitin Transformation

Similar to SAT, an interesting problem is CircuitSAT that asks whether a given Boolean circuit is satisfiable. The problem may be solved by reducing the CircuitSAT problem into an equisatisfiable SAT problem, where equisatisfiable means that the CNF formula will be satisfied only if the original circuit formula is satisfied. Tseitin transformation is used to generate CNF clauses for individual gates; the CNF formula for the overall circuit is a conjunction of the clauses generated for the individual gates.

Let us first consider logic gates on an individual basis. An AND gate with output $z = a \& b$ can be converted into an equisatisfiable CNF by realizing that $z \iff a \& b$. Here $x \iff y$ denotes that x and y are equivalent, which can also be written as $(x \implies y) \& (y \implies x)$, where $x \implies y$ denotes that x implies y. The implication $x \implies y$ is specified as $(!x + y)$ [2]. Thus,

$$
\begin{aligned}
z &\iff a \& b \\
&= (z \implies a \& b) \& (a \& b \implies z) \\
&= (!z + a \& b) \& (!(a \& b) + z) \\
&= (!z + a) \& (!z + b) \& (z + !a + !b) \\
&= (!z + a) (!z + b) (z + !a + !b)
\end{aligned}
\tag{4.1}
$$

The CNF formula for the AND, OR, INV, and XOR gates is specified in Fig. 4.1. Figure 4.2 shows a circuit and its CNF formula. The overall formula is obtained by ANDing the clauses for the individual gates.

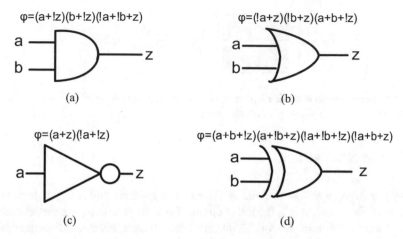

Fig. 4.1 CNF formulas for (**a**) AND, (**b**) OR, (**c**) INV, and (**d**) XOR gates, generated using the Tseitin transformation

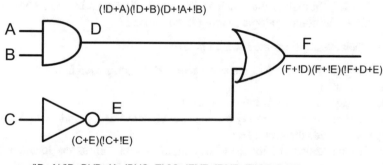

Fig. 4.2 Application of the Tseitin transformation to a circuit with three gates

4.1.3 Miter Circuit

An important use of SAT solvers is in equivalence checking, which is accomplished using a *miter* circuit. The concept of a miter circuit is essential for an in-depth understanding of the SAT attack operation. The miter circuit encodes equivalence checking for two Boolean circuits. As illustrated in Fig. 4.3, the two circuits have the same inputs. Each output of the two circuits is XORed, and the outputs of the XOR gates are ORed. The miter will be satisfiable (SAT), i.e., the OR gate will produce a 1 if any of the XOR gates produces a 1, indicating that the outputs of two circuits differ for an input pattern. The two circuits are equivalent if the miter is unsatisfiable (UNSAT), implying that the outputs of the two circuits match for all input patterns.

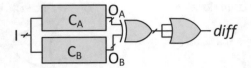

Fig. 4.3 An illustration of a miter circuit used for equivalence checking of Boolean circuits. The signal $diff = 1$ only if $O_A \neq O_B$ for any input pattern

4.2   The SAT Attack

In this section, we describe the SAT attack algorithm following a bottom-up approach. First, we explain the miter circuit that is used to compute distinguishing input patterns followed by a description of the attack algorithm. Beforehand, we state the threat model of the attack and present the notations used throughout the rest of this chapter.

Threat Model Similar to the sensitization attack discussed in Sect. 3.3, the SAT attack follows the standard logic locking threat model.

The attack assumptions are restated below:

- The designer is trusted, i.e., the personnel and the tools used in the design house are trustworthy.
- The foundry and the end-user are untrusted.
- The attacker has access to (1) a reverse-engineered netlist and (2) a functional chip that embeds the secret key.
- The attacker knows the logic locking algorithm, as well as the location of the protection circuitry. The only unknown is the secret key.

Notation The SAT attack operates using a locked netlist $L(I, K)$ and a functional IC $F(I)$, where $I \in \{0, 1\}^n$ denotes the primary inputs, and $K \in \{0, 1\}^k$ denotes the key inputs. The objective of the attacker is to retrieve a correct key K_C such that $L(I, K_C) = F(I)$, i.e., the locked netlist becomes functionally equivalent to the oracle.

4.2.1   Distinguishing Input Patterns (DIPs)

Similar to the miter circuit presented in Fig. 4.3, the SAT attack relies on a miter-like circuit that consists of two copies of the locked netlist. As illustrated in Fig. 4.4, the primary inputs are common to the two copies of the locked netlist, however, the key inputs K_A and K_B are left independent. The corresponding outputs of the two circuits are XORed and then ORed to generate the *diff* signal. As the CNF of the miter (generated using the Tseiten transformation) is passed to a SAT solver, the

Fig. 4.4 Miter circuit
employed by the SAT attack
to determine DIPs [10]

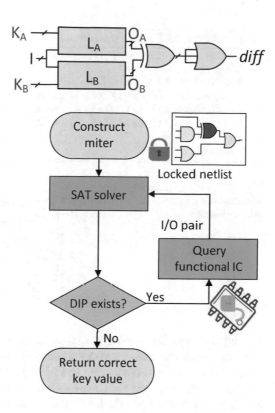

Fig. 4.5 Flowchart of the
SAT attack [10]

solver tries to find a satisfying assignment to the CNF formula with the constraint $diff = 1$. If the formula is satisfiable, the solver returns an input pattern I_d and two key values K_A and K_B. Since $diff = 1$, the outputs O_A and O_B of the two instances are different.

Suppose I_d is applied to the functional IC and the correct output O_d is obtained. Since O_A and O_B differ, at most one of the two key values (let's assume it is K_A) can generate an output consistent with O_d; the other key value (K_B) can be regarded as infeasible/incorrect since it produces an erroneous output for at least one input pattern. K_A can be categorized as a feasible/potentially correct key value. Since I_d classifies the key search space into classes, feasible and infeasible, it is referred to as a distinguishing input pattern. DIPs are at the core of the SAT attack; the attack is simply an iterative application of DIPs until no more DIPs can be found.

4.2.2 Attack Algorithm

As depicted in Fig. 4.5, the attack begins by constructing a miter-like circuit from the locked netlist. A SAT solver uses the CNF formula of the miter to compute a DIP I_d, which is then applied to the functional IC and the IC output O_d is recorded.

Algorithm 4: SAT attack algorithm [10]

Input   : Locked netlist $L(I, K)$, Functional IC $F(I)$
Output: Correct key K_C
1 **while** $I_d = SAT(L(I, K_A) \neq L(I, K_B))$ **do**
2 | $O_d = F(I_d)$ // Query the oracle
3 | $L(I, K_A) = L(I, K_A) \wedge (L(I_d, K_A) = O_d)$ // Augment clauses
4 | $L(I, K_B) = L(I, K_B) \wedge (L(I_d, K_B) = O_d)$
5 **end**
6 $K_C = SAT(L(I, K_A))$

Fig. 4.6 A locked circuit with three key and three primary inputs

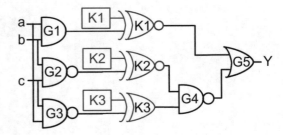

The input/output (I/O) pair (I_d, O_d) is added as a constraint to the SAT formula. The key values inconsistent with the constraint are thus eliminated from the search space. This process of computing and adding DIPs to the CNF formula is repeated and further incorrect key values are eliminated in subsequent attack iterations. The attack is successful when the SAT solver returns UNSAT, i.e., no further DIPs can be found, implying that all incorrect key values have been pruned. The SAT solver is then run once more without any constraint on the *diff* signal to extract a correct key K_C for the locked netlist. The construction of the SAT attack guarantees that it will retrieve a correct key upon completion.

The SAT attack is depicted in Algorithm 4. The miter along with the constraint on *diff* signal is represented as $L(I, K_A) \neq L(I, K_B)$. In each iteration, a DIP I_d is computed by finding a satisfying assignment to the miter CNF. The output O_d of the functional IC is recorded and based on the I/O pair (I_d, O_d), the CNF formula is augmented with more clauses. The attack terminates when the SAT solver returns UNSAT. The correct key values K_C is then computed by making the final call to SAT solver, which solves the formula $L(I, K_A)$.

Example Let us consider the application of the SAT attack on the logic locked circuit shown in Fig. 4.6. The circuit has three primary and three key inputs. Table 4.1 represents the output of the locked circuit for different key and input combinations. The values $(k0, \ldots, k7)$ represent all possible values for the three key inputs $\{K1, K2, K3\}$. The column Y depicts the correct circuit output, and each subsequent column depicts the output of the locked circuit for a specific key value.

In iteration 1, the attack randomly applies DIP 110, the corresponding correct output of the functional IC is 1. It can be observed that only $k2$ produces an incorrect output for the pattern 110; the remaining key values produce the correct

Table 4.1 SAT attack [10] on the locked circuit shown in Fig. 4.6

abc	Y	k0	k1	k2	k3	k4	k5	k6	k7	Incorrect keys identified
000	0	1	1	1	1	1	1	0	1	
001	0	1	1	1	1	1	1	0	1	
010	0	1	1	1	1	1	1	0	1	iter 3: other incorrect keys
011	1	1	1	1	1	0	1	1	1	
100	0	1	1	1	1	1	1	0	1	
101	1	1	1	1	1	1	1	1	0	
110	1	1	1	0	1	1	1	1	1	iter 1: k2
111	1	1	0	1	1	1	1	1	1	iter 2: k1

The red entries represent incorrect output generated by incorrect key values. $k6$ is the correct key value; the columns with all correct output values are shaded in blue

output. When the I/O pair (110, 1) is added to the SAT formula, the key value $k2$ is eliminated from the search space. The last column in the table lists the key values eliminated in each iteration. In the second iteration, the incorrect key value $k1$ is eliminated by applying the DIP 111. However, in iteration 3, the SAT solver returns the pattern 010, for which all key values except for $k6$ generate an erroneous output. Thus, the DIP 010 eliminates the remaining incorrect keys from the search space and the attack successfully identifies $k6$ as the correct key. Once the correct key value is retrieved, the attacker can use it to unlock/activate a non-functional IC. Another option is for the attacker to re-synthesize the locked netlist with the correct key to obtain the original netlist.

Note that the attack could have succeeded in the first iteration with a single DIP 010, if this input pattern was tried first. This indicates that the execution time of the attack depends on the order in which the SAT attack applies the input patterns. However, the SAT attack as well as its current variants [5], select the DIPs on a random basis [5, 10].

4.3 Effectiveness Against Pre-SAT Logic Locking

Table 4.1 also illustrates that each DIP may have a different distinguishing ability, which dictates how many keys values can be identified as incorrect by a given DIP. For example, the DIP 111 eliminates only one incorrect key whereas, the DIP 010 can eliminate seven incorrect keys. It is evident that if the attack applies more DIPs with high distinguishing ability, it can succeed with a fewer number of DIPs. However, as we pointed out earlier, the SAT attack applies DIPs on a random basis.

The ineffectiveness of the Pre-SAT logic locking techniques can be attributed to the fact that these techniques disregard the distinguishing ability of DIPs. Let us examine the techniques on an individual basis. As for FLL [6], it aims at maximizing the error injected into a circuit, inevitably enabling the SAT attack to generate DIPs with high distinguishing ability. In RLL [7], the key gates tend to be distributed uniformly across the netlist. Typically, most of the key gates are isolated from one another, enabling the SAT attack to work on small groups of key gates in a divide-and-conquer fashion. SLL tries to build cliques of key gates, leading to the localization of key gates in selected regions of a netlist and leaving the rest of the netlist unprotected. Consequently, the SAT attack needs to process a smaller number of clauses, resulting in a lower computational effort against SLL compared to either FLL or RLL. In summary, RLL, FLL, and SLL are easily broken by the SAT attack and there is a need for more powerful logic locking techniques. In the next section, we discuss the guiding principles for achieving a high SAT attack resilience.

4.4   How to Thwart the SAT Attack?

The total execution time of the SAT attack comprising λ iterations with t_i denoting the execution time for the ith iteration can be represented as:

$$T = \sum_{i=1}^{\lambda} t_i \tag{4.2}$$

According to Eq. 4.2, there are two potential ways of increasing the SAT attack resilience, represented in terms of the execution time of the attack:

1. Increase λ, the number of DIPs required for a successful SAT attack. This can be achieved by controlling the distinguishing ability of the DIPs. Techniques that thwart the SAT attack by rendering the number of DIPs exponential in the key size include SARLock [12], Anti-SAT [11], ATD [3], and SFLL [14]. SARLock [12], Anti-SAT [11], and ATD rely on point functions for controlling the distinguishing ability of DIPs and are referred to as "point function-based logic locking". SARLock, ATD, and Anti-SAT are presented in Chap. 5; SFLL is presented in Chap. 9.
2. Increase t_i, the time required for individual attack iterations. This task can be accomplished by making use of circuit structures that are inherently hard to resolve for the SAT solvers. Logic locking techniques that employ this strategy to thwart the SAT attack include ORF-Lock [13] and cyclic logic locking [8]. ORF-Lock makes use of one-way functions (see Sect. 8.3) whereas cyclic logic locking introduces dummy cycles/loops in a circuit (see Sect. 8.1).

4.5   Formal Security Analysis Framework

In this section, we lay down the foundations for formal security analysis of logic locking. We define terminology that will be used for security analysis in the subsequent chapters and the metrics used to quantify resilience against different attacks.

Notation An original circuit ckt_{orig} is a netlist that implements a Boolean function $F : I \rightarrow O$, where $I = \{0, 1\}^n$ and $O = \{0, 1\}^m$ with n inputs and m outputs. A logic locking technique $\mathcal{L}$ locks ckt_{orig} with the secret key k_{sec} to obtain the locked circuit ckt_{lock}. A locked circuit is activated using k_{sec} to obtain ckt_{act}. $\mathcal{A}^{\mathbb{S}}$ denotes an adversary $\mathcal{A}$ following an attack strategy $\mathbb{S}$. $\mathcal{A}^{\mathbb{S}}$ aims to recover a key k_{rec} or a netlist ckt_{rec} such that $ckt_{rec}(i) = ckt_{orig}(i), \forall i \in I$.

SAT Attack Resilience It is evident from the previous discussion that the SAT attack prunes the key search space iteratively by querying the oracle ckt_{act} with a DIP d_i in the ith iteration. The attack terminates upon querying the oracle with a selected set of DIPs, recovering a single key k_{rec}. Upon a successful attack, the recovered netlist ckt_{rec} is functionally equivalent to the original netlist ckt_{orig}, i.e., $ckt_{rec}(i) = ckt_{orig}(i), \forall i \in I$.

Definition 4.1 A logic locking technique $\mathcal{L}$ is called λ-secure against an adversary $\mathcal{A}^{\mathrm{SAT}}$ making a polynomial number of queries $q(\lambda)$ to the oracle, if he/she cannot reconstruct ckt_{rec} correctly with the probability P_{succ} greater than $\frac{q(\lambda)}{2^\lambda}$.

For the basic logic locking techniques, which are oblivious to the distinguishing ability of DIPs, $\mathcal{A}^{\mathrm{SAT}}$ can extract ckt_{rec} correctly, using only a polynomial number of queries $q(|k|)$ with $|k|$ denoting the key size, as the probability $P_{succ} \gg \frac{q(|k|)}{2^{|k|}}$.

Sensitization Attack Resilience The resilience against the sensitization attack is defined in terms of the number of *pairwise-secure* key gates.

Definition 4.2 A logic locking technique $\mathcal{L}$ is λ-secure against a sensitization attack iff λ key bits are all pairwise secure.

Approximate Attack Resilience The resilience against an approximate attack is defined in terms of the OER between the original netlist output and the locked netlist output obtained by applying random incorrect key values to the locked netlist.

In summary, this chapter presented the SAT attack, the associated threat model, the attack algorithm, the effectiveness against Pre-SAT logic locking techniques, and potential ways to thwart the SAT attack. The SAT attack has steered the course of logic locking research; the recent research efforts focus on developing cost-effective and provably-secure countermeasures against the SAT attack. Chapter 5 presents point-function based logic locking techniques, which are amongst the earliest countermeasures against the attack.

References

1. Biere A (2013) Lingeling, plingeling and treengeling entering the sat competition 2013. SAT Competition, pp 51–52
2. Ikenaga B (2009) Truth tables, tautologies, and logical equivalence. http://sites.millersville. edu/bikenaga/math-proof/truth-tables/truth-tables.html

3. Li M, Shamsi K, Meade T, Zhao Z, Yu B, Jin Y, Pan D (2016) Provably secure camouflaging strategy for IC protection. In: IEEE/ACM international conference on computer-aided design, pp 28:1–28:8
4. Liu D, Yu C, Zhang X, Holcomb D (2016) Oracle-guided incremental SAT solving to reverse engineer camouflaged logic circuits. In: Design, automation test in Europe, pp 433–438
5. Massad M, Garg S, Tripunitara M (2015) Integrated circuit (IC) decamouflaging: reverse engineering camouflaged ICs within minutes. In: Network and distributed system security symposium
6. Rajendran J, Zhang H, Zhang C, Rose G, Pino Y, Sinanoglu O, Karri R (2015) Fault analysis-based logic encryption. IEEE Trans Comput 64(2):410–424
7. Roy JA, Koushanfar F, Igor L (2008) EPIC: ending piracy of integrated circuits. In: Design, automation, and test in Europe, pp 1069–1074
8. Shamsi K, Li M, Meade T, Zhao Z, Pan DZ, Jin Y (2017) Cyclic obfuscation for creating sat-unresolvable circuits. In: ACM Great Lakes symposium on VLSI, pp 173–178
9. Sorensson N, Een N (2005) Minisat v1. 13-A sat solver with conflict-clause minimization. SAT 2005(53):1–2
10. Subramanyan P, Ray S, Malik S (2015) Evaluating the security of logic encryption algorithms. In: IEEE international symposium on hardware oriented security and trust, pp 137–143
11. Xie Y, Srivastava A (2016) Mitigating SAT attack on logic locking. In: International conference on cryptographic hardware and embedded systems, pp 127–146
12. Yasin M, Mazumdar B, Rajendran J, Sinanoglu O (2016) SARLock: SAT attack resistant logic locking. In: IEEE international symposium on hardware oriented security and trust, pp 236–241
13. Yasin M, Rajendran J, Sinanoglu O, Karri R (2016) On improving the security of logic locking. IEEE Trans Comput Aided Des Integr Circuits Syst 35(9):1411–1424
14. Yasin M, Sengupta A, Nabeel MT, Ashraf M, Rajendran J, Sinanoglu O (2017) Provably-secure logic locking: from theory to practice. In: ACM/SIGSAC conference on computer & communications security, pp 1601–1618

Chapter 5
Post-SAT 1: Point Function-Based Logic Locking

Abstract This chapter presents point function-based logic locking techniques, namely SARLock, Anti-SAT, and AND-tree detection that thwart the SAT attack by controlling the distinguishing ability of the DIPs. All these techniques integrate with the original netlist a point function that sets a limit on the number of incorrect key values that a DIP can eliminate. While these techniques cost-effectively thwart the SAT attack, their main limitation is the susceptibility to removal attacks. Moreover, these techniques fail to achieve a high output error rate.

This chapter is about point function-based logic locking techniques that remain the earliest countermeasures against the SAT attack. The chapter presents three techniques, SARLock [5], Anti-SAT [4], and AND-tree detection [1]. All three techniques harness point functions to control the distinguishing ability of individual DIPs and ultimately render the required number of required DIPs exponential in the key size. Section 5.1 describes the common principle underlying these techniques. Section 5.2 describes the architecture and operation of SARLock. Section 5.3 explains how Anti-SAT makes use of point functions to circumvent the SAT attack. Section 5.4 elaborates on the effectiveness of AND-tree detection. Section 5.5 compares the three approaches in terms of the attack resilience and the implementation cost. Section 5.6 highlights the limitations of the point function-based logic locking.

5.1 Maximizing SAT Attack Resilience

5.1.1 Strong and Weak DIPs

From the discussion in Sect. 4.4, it is evident that resilience against the SAT attack can be achieved by controlling the distinguishing ability of individual DIPs. While a *strong* DIP can eliminate a relatively large number ($\gg 1$) of incorrect key values,

© Springer Nature Switzerland AG 2020
M. Yasin et al., *Trustworthy Hardware Design: Combinational Logic
Locking Techniques*, Analog Circuits and Signal Processing,
https://doi.org/10.1007/978-3-030-15334-2_5

Table 5.1 Maximal
resilience against the SAT
attack can be achieved by
using weak DIPs

abc	Y	k0	k1	k2	k3	k4	k5	k6	k7
000	0	0	0	0	0	0	0	0	1
001	0	0	0	0	0	1	0	0	0
010	0	0	0	0	1	0	0	0	0
011	1	1	1	1	1	1	1	1	0
100	0	1	0	0	0	0	0	0	0
101	1	1	1	1	1	1	1	1	1
110	1	1	0	1	1	1	1	1	1
111	1	1	1	0	1	1	1	1	1

At most one incorrect key value corrupts the output
for any DIP. The number of DIPs required by the SAT
attack is $2^3 - 1 = 7$

a *weak* DIP can eliminate only a few key values ($\approx$1). Thus, maximal resilience
against the SAT attack can be obtained by ensuring that a logic locking technique is
subject to only weak DIPs. This basic principle is at the core of all point function-
based logic locking techniques that exhibit strong resilience against the SAT attack.
The Pre-SAT logic locking techniques, however, tend to is subject to strong DIPs,
resulting in minimal resilience against the SAT attack.

Example Table 5.1 illustrates the worst-case scenario for the SAT attack, where the
attack can eliminate at most one incorrect key value per DIP. It can be observed
that in each row, there is at most one key value that generates an incorrect output,
resulting in weak DIPs. When the SAT attack is launched on a locked circuit that
exhibits a behavior similar to that depicted in the table, the required number of DIPs
is exponential in the key size k, i.e., $\#DIPs = 2^k - 1$ in the worst case.

5.1.2   Circuits that Generate Weak DIPs

So far, we have established that one option to achieve high SAT attack resilience is to
force the generation of weak DIPs. A question that arises is what kind of circuits can
generate such DIPs. To answer that question, let us consider the individual columns
in Table 5.1. We observe that in the column $k2$, there is only red (incorrect) entry for
the DIP 111. By concatenating all the entries in the column, we obtain the Boolean
vector 00010110. By XORing the column vector Y with the column vector $k2$, we
obtain 00010110 $\oplus$ 00010111 = 00000001. The resulting vector contains a single
1; the remaining entries in the vector are all 0s. Similar vectors, containing single
1s, can be obtained by XORing the column vector Y with the other column vectors.
As already discussed, point functions output a 1 for only one input value. We note
that each column in the table can be realized using a circuit that implements a point
function, e.g., an AND gate or a NOR gate. As illustrated in Fig. 5.1, point function-
based logic locking techniques integrate a point function with the original circuit

Fig. 5.1 The architecture of point function-based logic locking techniques [5]

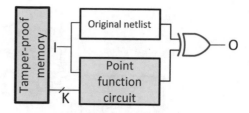

through an XOR gate. We elaborate on this aspect in the forthcoming sections of this chapter, which describe how specific logic locking techniques make use of point functions to thwart the SAT attack.

5.2 SARLock

SARLock, which abbreviates, "SAT Attack Resistant Logic Locking," is the first technique that exploits point functions to resist the SAT attack in a scalable and cost-effective manner [5]. SARLock also achieves SAT attack resilience by restraining the distinguishing ability of DIPs. In this section, we describe the architecture of SARLock followed by security analysis against SAT and other attacks.

5.2.1 Architecture

As illustrated in Fig. 5.2, SARLock integrates a comparator with the original circuit. The comparator is essentially an AND gate (a point function) with an XNOR key gate inserted at each of its inputs. The output of the comparator is the *flip* signal that is asserted whenever the key inputs match the primary inputs, i.e., along the diagonal in Table 5.2. The *flip* signal is XORed with one of the primary outputs of the circuit and injects a controllable error (for only one DIP corresponding to an incorrect key) into the circuit. To prevent the *flip* signal from being asserted for the correct key value—$k6$ in this example—a small mask logic is inserted. The mask circuit hard-codes the secret key value k_s using a large k-input AND/NOR gate and nullifies the error introduced by the comparator. Assuming $F(I)$ denotes the original circuit, the output O of the locked circuit can be represented as $O = F(I) \oplus ((I == K)\&(I \neq k_s))$, where K denotes the key inputs. In Table 5.2, each input pattern is a weak DIP and can eliminate at most one incorrect key, yielding the required number of DIPs to be an exponential function of key size.

For a key size k, the SARLock protection circuitry consists of $k + 1$ 2-input XOR/XNOR gates and $2k + 1$ 2-input AND gates. On increasing the key size k, the area overhead grows linearly, while there is an exponential increase in the #DIPs.

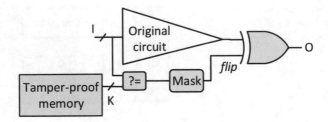

Fig. 5.2 The SARLock circuit to resist the SAT attack. The *flip* signal is asserted upon a match between an input value and a key value. The Mask circuit restores the correct output for the secret key value

Table 5.2 SARLock resilience against the SAT attack

abc	O	k0	k1	k2	k3	k4	k5	k6	k7
000	0	1	0	0	0	0	0	0	0
001	0	0	1	0	0	0	0	0	0
010	0	0	0	1	0	0	0	0	0
011	1	1	1	1	0	1	1	1	1
100	0	0	0	0	0	1	0	0	0
101	1	1	1	1	1	1	0	1	1
110	1	1	1	1	1	1	1	1	1
111	1	1	1	1	1	1	1	1	0

All incorrect entries occur across the diagonal. The output is always correct for the correct key value $k6$

5.2.2 Security Analysis

In this section, we provide theoretical proofs for the security of SARLock against different attacks.

SAT Attack Resilience As already pointed out, SARLock achieves high SAT attack resilience by using weak DIPs that eliminate at most one incorrect key per iteration of the attack.

Theorem 5.1 *SARLock is k-secure against the SAT attack.*

Proof There exist a total of $2^k - 1$ incorrect key values, with k denoting the key size. Using a single DIP, the SAT attack can eliminate exactly one incorrect key value, leaving $2^k - 2$ incorrect values still to be eliminated. The probability of recovering the correct netlist is only $\frac{1}{2^k-1}$. With q queries, only q incorrect key candidates are eliminated, and the probability of recovering the correct netlist $P_{succ} \approx \frac{q}{2^k}$, implying that SARLock is k-secure against the SAT attack.

Sensitization Attack Resilience Recall that sensitization attack aims at sensitizing individual key bits to primary outputs. The attack can be thwarted by inserting key gates that protect one another.

Theorem 5.2 *SARLock is k-secure against the sensitization attack.*

Proof In SARLock, all the k bits of the key converge within the comparator to produce the *flip* signal. Therefore, sensitizing any key bit through the *flip* signal to the output O requires controlling all the other key bits. All k bits are therefore pairwise-secure, implying that SARLock is k-secure against sensitization attack.

Removal Attack Resilience While SARLock is secure against the SAT and sensitization attacks, it cannot withstand the removal attacks. Since the protection logic, comprising the comparator and the Mask blocks, is isolated and not intertwined/merged with the original circuit, an attacker can easily separate it from the original circuit. For more discussion on removal attacks, refer to Chap. 7.

Approximate Attack Resilience SARLock is a low-corruptibility technique since the error injected by SARLock into a circuit is minimal. An attacker can apply an arbitrary key value to an IC, and still, recover the correct output for most of the input patterns. Table 5.2 demonstrates that for an arbitrary incorrect key value, only one input pattern will generate an incorrect output, resulting in an OER of $\frac{1}{2^k}$.

5.3 Anti-SAT

The basic idea behind Anti-SAT is to use complementary Boolean functions, such as AND and NAND, to achieve controllable resilience against the SAT attack [4]. By using two complementary functions, Anti-SAT avoids the need to hard-code the secret, which is required by SARLock. While SARLock has k key inputs, the Anti-SAT has $2k$ key inputs.

5.3.1 Architecture

The Anti-SAT block comprises two blocks, $B_1 = g(X, K_{l1})$ and $B_2 = \overline{g(X, K_{l2})}$, as shown in Fig. 5.3 [4]. The two blocks share the same inputs X but have different key inputs K_{l1} and K_{l2}. The outputs of B_1 and B_2 drive an AND gate to produce the output signal Y. The two blocks produce complementary outputs when the correct key value is applied; $Y = 0$ for all inputs, leading to no error injection, and thus, to a correct output. For an incorrect key value, the outputs of both B_1 and B_2 are 1 for a specific input pattern; for that pattern, $Y = 1$, leading to the injection of an error. Assuming that Anti-SAT protects one of the primary outputs of the original circuit $F(I)$, the protected output O can be represented as $O = F(I) \oplus (g(X \oplus K_{l1}) \wedge g(X \oplus K_{l2}))$.

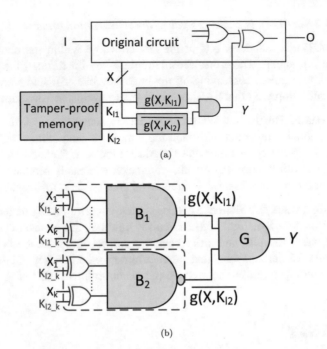

Fig. 5.3 (a) Integration of Anti-SAT with an original circuit, and (b) an instance of Anti-SAT that provides the maximum resilience against the SAT attack [4]

5.3.2 Security Analysis

The security properties of Anti-SAT are dictated by p, the on-set of the function g [4]. Anti-SAT typically uses a value of p close to either 1 or $2^k - 1$. Note that $p = 1$ for the Anti-SAT circuit shown in Fig. 5.3b.

SAT Attack Resilience The basic intuition for high SAT attack resilience of Anti-SAT is the same as that for SARLock, i.e., the use of weak DIPs that can eliminate only a few incorrect key values. The main difference is that Anti-SAT can vary p to adjust the security level against the SAT attack. Taking p into account, the lower bound on the number of DIPs required for a successful SAT attack against Anti-SAT can be represented as (for the detailed proof refer to [4]):

$$\#DIPs = \frac{2^{2k} - 2^k}{p(2^k - p)}. \tag{5.1}$$

For $p \in \{1, 2^k - 1\}$, $\#DIPs \approx 2^k$, i.e., exponential in the key size.

Sensitization Attack Resilience Similar to the case of SARLock, all the $2k$ key bits in Anti-SAT converge to produce the output signal Y. All $2k$ key bits are pairwise secure, implying that Anti-SAT is $2k$-secure against the sensitization attack.

Removal Attack Resilience   In Anti-SAT, the original circuit is implemented as is without any changes. The symmetric construction of Anti-SAT allows an attacker to identify, and consequently, remove the Anti-SAT block. A detailed description of removal attacks on Anti-SAT is presented in Chap. 7.

Approximate Attack Resilience   In Anti-SAT, the error injected into a locked circuit is dictated by p. For an arbitrary incorrect key value, around p input patterns will generate erroneous output, hinting at an expected OER of $\frac{p}{2^k}$.

5.3.3   Functional and Structural Obfuscation

Xie et al. [4] anticipated that a trivial attack could simulate the basic Anti-SAT netlist and find the complementary pair of signal outputs of g and $\bar{g}$, leading to the identification and removal of the Anti-SAT block. To prevent such simple removal attacks, k additional XOR/XNOR key gates may be inserted randomly at the inputs of the Anti-SAT block, obscuring the complementary relations between the signals, thereby, providing *functional obfuscation*.

Another simple attack could be in the form of circuit partitioning to identify the isolated Anti-SAT block and remove it from the netlist [4]. To thwart such attacks, *structural obfuscation* using MUX-based logic locking has been proposed to increase the inter-connectivity between the logic locked circuit and the basic Anti-SAT (BA) block [4]. The resultant obfuscated Anti-SAT (OA) block has $4n$ key gates. Figure 5.4 illustrates an obfuscated Anti-SAT block. In Chap. 7, we highlight on the security vulnerabilities of obfuscated Anti-SAT and show how it remains vulnerable to more sophisticated forms of removal attacks.

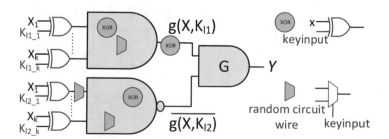

Fig. 5.4 Functional and structural obfuscation in Anti-SAT. The XOR key gates alter the signal probabilities and break the symmetry of the signals. The MUX key gates obscure the boundary between Anti-SAT and the original circuit [4]

5.4 AND-Tree Detection

The underlying principle of ATD is the same as that of SARLock and Anti-SAT, i.e., utilize point functions to control the distinguishing ability of DIPs. However, as opposed to Anti-SAT and SARLock, which integrate external point functions with the original netlist, ATD identifies such structures inside an original netlist and reuses them in order to decrease the implementation cost [1]. Once a point function, e.g. an AND-tree or an OR-tree, has been identified in a netlist, ATD locks the tree by inserting XOR/XNOR key gates at the inputs of the tree. An existing AND-tree can be identified by running a breadth-first search on the circuit graph and tracking the types (AND/OR/NAND/NOT etc.) of the gates in the circuit [1]. Note that the tree input may not be the primary inputs of the circuits. Figure 5.5 shows a logic locked AND-tree.

When the foundry is trusted and protection is required only against malicious end-users, the tree inputs may be camouflaged by inserting INV/BUF camouflaged gates at each of the tree input. Figure 5.5 also shows the camouflaging counterpart of the locked AND-tree.

5.4.1 Security Analysis

The security of ATD is dictated by the size of the *largest non-decomposable* tree in the circuit, i.e., a tree where all internal nodes have a fan-out of one. If the internal nodes of a tree have multiple fan-outs, an attacker can partition the tree into subtrees and target the subtrees on an individual basis. An example non-decomposable AND-tree and a decomposable tree is presented in Fig. 5.6a, b, respectively. To achieve sufficient security against the SAT attack, large non-decomposable AND/OR-trees,

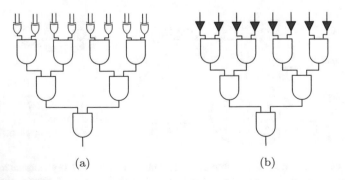

(a)   (b)

Fig. 5.5 (**a**) A locked AND-tree with XOR/XNOR key gates inserted at its inputs, and (**b**) the camouflaged counterpart of the AND-tree with camouflaged INV/BUF gates inserted at its inputs [1]. Both trees achieve the same level of security against the SAT attack (under the associated threat model)

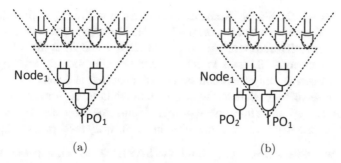

Fig. 5.6 (a) A non-decomposable AND-tree, and (b) a decomposable AND-tree [1]. Attacks on the decomposable tree can leverage divide-and-conquer strategies

Table 5.3 SAT attack resilience of ATD [1] for a 3-input AND gate with XOR key gates inserted at the tree inputs; the correct key is $k0$

abc	O	k0	k1	k2	k3	k4	k5	k6	k7
000	0	0	0	0	0	0	0	0	1
001	0	0	0	0	0	0	0	1	0
010	0	0	0	0	0	0	1	0	0
011	0	0	0	0	0	1	0	0	0
100	0	0	0	0	1	0	0	0	0
101	0	0	0	1	0	0	0	0	0
110	0	0	1	0	0	0	0	0	0
111	1	1	0	0	0	0	0	0	0

For any given DIP except for the pattern 111, the SAT attack can eliminate at most one incorrect key value, i.e., the one that injects an error at the output. The pattern 111, if applied, will eliminate all the incorrect key values in a single iteration

e.g., with 64 or 128 inputs, are required. Such large trees are rare in common benchmark circuits. We elaborate on this in Sect. 7.3. For the upcoming discussion in this section, we assume that non-decomposable AND-trees exist in a netlist. Furthermore, we assume that all the tree inputs are primary inputs, representing the best possible security achievable by ATD.

SAT Attack Resilience Table 5.3 presents the output of a 3-input AND-tree locked using ATD for different key and input combinations. It can be observed that exactly one incorrect key value can be eliminated by any of the input patterns, except for one special input pattern 111, which if applied, can identify all incorrect keys. If an attacker somehow identifies and applies this specific DIP, the attack will succeed in a single iteration. However, currently, there exists no known algorithm that can identify such DIPs from the analysis of a locked netlist. The average number of patterns that an attacker is expected to exercise (on a random basis) prior to exercising the special input pattern is $2^{(k-1)}$. The SAT attack resilience of ATD grows exponentially with increasing key size, similar to that for SARLock and Anti-SAT.

Removal Attack Resilience Removal attacks are not directly applicable to ATD since the AND-tree is part of the original netlist. However, as will be illustrated in Sect. 7.3, large AND-trees rarely exist in real circuits and ATD often has to make use of large dummy AND-trees to achieve a reasonable security level against the SAT attack. Such dummy trees can be easily identified and removed, effectively reducing the security of ATD to the size of real AND-trees. Another weakness of ATD is that the tree inputs may not be primary inputs; the associated bias in the input distribution of the tree leads to a reduction in the security level (refer to Sect. 7.3).

Approximate Attack Resilience Similar to SARLock, ATD is a low-corruptibility technique exhibiting an OER of $\frac{2}{2^k}$. An attacker can apply an arbitrary key value to a ATD-locked IC, and still recover the correct output for most of the input patterns.

5.5   A Comparative Analysis

Table 5.4 presents a comparison of the point function-based logic locking techniques in terms of their computational effort (#DIPs) and the implementation overhead. In terms of the computational effort, the #DIPs required by SARLock and Anti-SAT is $2^k - 1$. For ATD, however, an attacker may be fortuitous and succeed even upon applying one input pattern. The average #DIPs for the ATD is still exponential in the key size, i.e., 2^{k-1}.

Among the three techniques, Anti-SAT incurs the highest implementation overhead as it comprises of a k-input AND-tree and a k-input NAND-tree. The overhead of ATD is the lowest as it basically requires k XOR gates if an AND-tree is detected in the netlist. However, if a tree of sufficient size is not found, a dummy AND-tree may be inserted, resulting in an increased overhead.

5.6   The Common Pitfalls

As we have already pointed out throughout this chapter, the point function-based logic locking exhibits two main drawbacks.

Table 5.4 A comparison of the point function-based logic locking techniques

Technique	Computational effort (#DIPs)			Implementation overhead
	Best	Avg.	Worst	
SARLock [5]	$2^k - 1$	$2^k - 1$	$2^k - 1$	k-bit AND-tree + k-bit comparator
Anti-SAT [4]	$2^k - 1$	$2^k - 1$	$2^k - 1$	k-bit AND-tree + k-bit NAND-tree + $2k$ XORs
ATD [1]	$2^k - 1$	$2^{(k-1)}$	1	k-bit AND-tree + k XORs

It is assumed that $p = 1$ in Anti-SAT

1. **Low OER**. The SAT attack resilience of all point function-based locking techniques stems from the controlled error injection into a circuit. The higher the injected error, the lower the SAT attack resilience. Consequently, these techniques exhibit very small OER. Even for arbitrary key values, an attacker may be able to retrieve the correct circuit output for a large fraction of input patterns. Integration of these low-corruptibility techniques with high corruptibility techniques such as RLL, FLL, etc. remains susceptible to approximate attacks, which can reduce the compound logic locking problem to its low-corruptibility counterpart [2, 3] (refer to Chap. 6 for a description of various approximate attacks).

2. **Structural traces**. The unique properties of point functions render them easily identifiable from the typical circuit structures. Consequently, each point function-based logic locking technique has associated with it specific structural traces, making it vulnerable to removal attacks (refer to Chap. 7 for discussion on removal attacks).

In summary, this chapter presented point function-based logic locking techniques, i.e., SARLock, Anti-SAT, and ATD. These techniques successfully thwart the SAT attack; however, they exhibit low output corruptibility and remain susceptible to removal attacks. The next chapter discusses approximate attacks that target compound logic locking techniques.

References

1. Li M, Shamsi K, Meade T, Zhao Z, Yu B, Jin Y, Pan D (2016) Provably secure camouflaging strategy for IC protection. In: IEEE/ACM international conference on computer-aided design, pp 28:1–28:8
2. Shamsi K, Li M, Meade T, Zhao Z, Pan DZ, Jin Y (2017) AppSAT: approximately deobfuscating integrated circuits. In: IEEE international symposium on hardware oriented security and trust, pp 95–100
3. Shen Y, Zhou H (2017) Double DIP: re-evaluating security of logic encryption algorithms. Cryptology ePrint Archive, Report 2017/290, http://eprint.iacr.org/2017/290
4. Xie Y, Srivastava A (2016) Mitigating SAT attack on logic locking. In: International conference on cryptographic hardware and embedded systems, pp 127–146
5. Yasin M, Mazumdar B, Rajendran J, Sinanoglu O (2016) SARLock: SAT attack resistant logic locking. In: IEEE international symposium on hardware oriented security and trust, pp 236–241

Chapter 6
Approximate Attacks

Abstract This chapter presents approximate attacks on logic locking, namely AppSAT and Double-DIP. Approximate attacks target compound logic techniques and reduce a compound technique (comprising a low corruptibility technique and a high corruptibility technique) to the low corruptibility technique. AppSAT augments the basic SAT attack with random queries at regular intervals; it terminates when the computed OER is below a certain threshold. Double-DIP makes use 2-DIPs that eliminate at least two incorrect key values per 2-DIP; the attack terminates when 2-DIPs can no longer be found. The effectiveness of the approximate attacks cautions against naive integration of logic locking techniques.

This chapter is about approximate attacks on logic locking. Approximate attacks target compound logic locking techniques, reducing the compound technique to its constituent low-corruptibility technique. Section 6.1 introduces compound logic locking techniques that paved the way for the emergence of approximate attacks. Section 6.2 elaborates on the operation of the AppSAT attack [3]. Section 6.3 describes how Double-DIP attack [4] works and how the attack methodology differs from that of the AppSAT attack.

6.1 Introduction

6.1.1 Compound Logic Locking

As mentioned in Sect. 5.6, a common drawback of point function-based logic locking techniques is their low output corruptibility. In an attempt to improve the output corruptibility, researchers have proposed compound logic locking techniques that combine a low output corruptibility technique (e.g., SARLock or Anti-SAT) with a high output corruptibility technique (e.g. RLL [2] or FLL [1]). For example, in [7], SARLock is integrated with SLL, the overall compound technique being

© Springer Nature Switzerland AG 2020
M. Yasin et al., *Trustworthy Hardware Design: Combinational Logic Locking Techniques*, Analog Circuits and Signal Processing,
https://doi.org/10.1007/978-3-030-15334-2_6

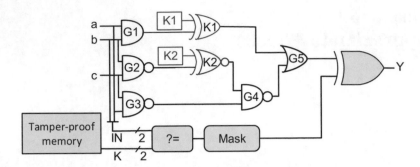

Fig. 6.1 An example of compound logic locking. A circuit locked using 2-bit RLL and 2-bit SARLock. The correct key is $k8$

Table 6.1 Output of a SARLock + RLL circuit for different key and input combinations

abc	O	k0	k1	k2	k3	k4	k5	k6	k7	k8	k9	k10	k11	k12	k13	k14	k15
000	0	1	1	1	1	1	1	1	1	0	0	0	0	1	1	1	1
001	0	1	1	1	1	1	1	1	1	0	0	0	0	1	1	1	1
010	0	1	0	1	1	1	0	1	1	0	1	0	0	1	0	1	1
011	1	1	0	1	1	1	0	1	1	1	0	1	1	0	1	0	0
100	0	1	1	0	1	1	1	0	1	0	0	1	0	1	1	0	1
101	1	1	1	0	1	1	1	0	1	1	1	0	1	1	1	0	1
110	1	0	0	0	1	1	1	1	0	1	1	1	0	1	1	1	0
111	1	1	1	1	0	1	1	1	0	1	1	1	0	1	1	1	0

The correct key value is $k8$

SARLock + SLL. The original Anti-SAT paper proposes to lock the original circuit with FLL and then integrate the locked circuit with the Anti-SAT block [6]. The low output corruptibility technique hampers the SAT attack, and the high corruptibility scheme improves the OER.

Example Let us again consider the majority circuit, presented earlier in Fig. 1.7a and lock it using (a) 2-bit RLL and (b) 2-bit SARLock. As illustrated in Fig. 6.1, the SARLock+RLL circuit has four key inputs. Table 6.1 presents the output of the locked circuit for different input and key combinations. It is evident that the number of incorrect entries in each column of the table is much higher compared to that for SARLock, presented in Table 5.2, where there is only one incorrect entry in each column.

6.1.2 Approximate Attacks

Most of the existing logic locking attacks are "exact" in nature, in the sense that the key value k_{rec} recovered by an attack renders the extracted netlist L functionally equivalent to the original netlist. The outputs for the two netlists match for all input patterns.

$$\underset{i \in I}{\forall} \, L(i, k_{rec}) = F(i) \tag{6.1}$$

As opposed to the exact attacks, the approximate attacks aim at recovering a key value that renders the recovered netlist "approximately" the same as the original netlist. The output of the two netlists may vary but for only a small subset of the input patterns, i.e., the OER of the recovered netlist is low. Putting it another way, the probability P_m of a mismatch between the outputs of the two circuits is negligibly small [3].

$$P_m = \underset{i \in I, k \in K}{Pr} \, L(i, k) \neq F(i) \tag{6.2}$$

For a logic locking technique (e.g., SARLock + RLL) characterized by $P_m = 0.25$, the output of the recovered and the original netlist will mismatch for one in four input patterns. The objective of an approximate attack is to recover a key value that nullifies the impact of RLL and renders P_m exponentially small, i.e., $\approx \frac{1}{2^k}$, which is a characteristic of the point function-based logic locking techniques.

Approximate attacks are effective when an attacker is not interested in recovering an exact netlist and it suffices for him/her to extract only a partially working netlist by expending relatively low attack effort. Let us consider that SARLock + RLL is applied to a microprocessor such that SARLock corrupts the microprocessor output for a small subset of the instructions [9] and RLL serves to increase the output corruptibility. By reducing SARLock + RLL to SARLock, an attacker can obtain a microprocessor that functions correctly for all but a few instructions and meets the goals of the attacker. Thus, approximate attack target a compound logic locking technique and reduce it to a low-corruptibility technique [3]. The threat model for these attacks is the same as that of the SAT attack, i.e., the attacker has access to (1) a reverse-engineered netlist and (2) a functional IC. While these attacks produce only an approximate netlist, they can be used as a pre-processing step for more sophisticated attacks that peel off defenses one at a time and ultimately retrieve an exact netlist. We elaborate on this in Sects. 7.2 and 7.4.

6.2 AppSAT

6.2.1 Basic Idea

The AppSAT attack algorithm builds on the top of the SAT attack. AppSAT relies on the observation that the SAT attack can be considered as an *active learning* problem, which is a special case of semi-supervised machine learning [3]. The objective of the attacker is to learn the target function F, within the hypothesis space C, by querying an oracle (functional IC). The subset of the hypothesis space that is consistent with the current set of queries (DIPs) is called the version space V (set of possible functions for the locked netlist, or equivalently the set of valid key values).

Among the active-learning strategies, the SAT attack most resembles the *query-by-disagreement strategy*. By querying the oracle with inputs that lead to a mismatch between functions in the version space, the attack keeps narrowing down the size of the version space until all functions in V are equivalent to the F. Recall that SAT attack is an exact attack; it terminates when no further disagreeing inputs can be determined. In the worst case, it may need to exercise an exponential number of queries. AppSAT, on the other hand, relaxes the exactness constraint and terminates much earlier compared to the SAT attack. AppSAT aims at guaranteeing an upper bound (ε) on the error rate of the recovered netlist and ensures that $P_m < \varepsilon$. Since computing an exact P_m is impractical, AppSAT resorts to computing only an estimate. Furthermore, to reduce the execution time of the algorithm, AppSAT makes use of *random query enforcement* [3]. We elaborate on these concepts in the subsequent sections.

6.2.2   Termination Criterion

The main difference between the SAT attack and AppSAT is the termination criterion. AppSAT terminates when P_m is below the threshold ε, typically set as $\frac{1}{2^k}$. Since computing the exact P_m requires an exponential number of queries to the oracle, AppSAT resorts to computing only an estimate of P_m. AppSAT makes q queries to the oracle after every d iterations of the attack. The values of q and d are set empirically to 50 and 12, respectively. It is assumed that 50 queries are sufficient to estimate the error for compound logic techniques (e.g. SARLock + RLL), since the low-corruptibility technique (i.e., SARLock) will produce an error for only one in 2^k input patterns [3]. In the initial iterations of the attack, the influence of high output corruptibility technique, i.e. RLL, will be dominant on the output, leading to a high P_m. Over successive attack iterations, the RLL key values will be resolved and P_m will gradually decrease. The attack terminates when $P_m < \varepsilon$. The experiments conducted by [8] suggest that 50 input patterns may be insufficient to estimate P_m and 1000 or more input patterns may be used to compute a better estimate.

6.2.3   Random Query Enforcement

Since AppSAT consults with the oracle with q queries after every d iterations, the oracle output for the q queries is already available. It has been empirically established that the adding these I/O pairs as new constraints to the SAT formula helps speed up the execution time of the AppSAT attack. This addition of random queries to speed up the attack is referred to random query enforcement.

Fig. 6.2 Flowchart of the
AppSAT attack [3]

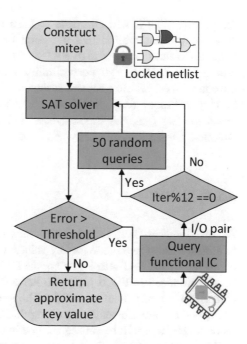

6.2.4   Attack Algorithm

As discussed earlier, AppSAT builds upon the SAT attack by querying the functional
IC with a fixed number of random DIPs at regular intervals and augmenting the CNF
formula with new constraints based on these DIPs. Moreover, the main difference
from the SAT attack is the termination criterion. As illustrated in Fig. 6.2, AppSAT
terminates when the estimated error (P_m) is below the specified threshold (ε).
Upon termination, the attack returns an *approximate key*. If the locked netlist is
re-synthesized using the extracted key value as a constraint on the key inputs, an
approximate netlist will be recovered [3].

The attack cautions against the naive integration of multiple logic locking
techniques; the overall design may still be insecure.

6.3   Double-DIP

6.3.1   Basic Idea

Similar to AppSAT, Double-DIP is an approximate attack that aims at reducing
a compound logic locking technique to its low-corruptibility component [4]. The
attack methodology and termination criterion, however, are different. At the heart
of Double-DIP is a specific class of DIPs, referred to as 2-DIPs; each 2-DIP has

the property that it will eliminate *at least* two incorrect key values. Double-DIP relies on the observation that the point function-based logic locking techniques (e.g., SARLock) achieve high SAT attack resilience by introducing only 1-DIPs into a circuit. The high corruptibility locking techniques (e.g., RLL) tend to introduce DIPs that eliminate more than one key value. Thus, when compound logic locking technique (e.g., SARLock + RLL) has been reduced the point function-based locking (i.e., SARLock), there will exist only 1-DIPs in the circuit. The Double-DIP attack terminates when 2-DIPs can no longer be found.

6.3.2   2-DIPs

As mentioned earlier, Double-DIP relies on 2-DIPs, which have the ability to eliminate at least two incorrect key values. Let us reconsider the basic miter circuit employed by the SAT attack to compute DIPs; it was introduced in Chap. 4 and has been reproduced in Fig. 6.3a. The miter XORs the outputs of the two circuits and then ORes the outputs of the XORs. The miter output is a 1 only when the outputs of the two circuits mismatch, ensuring that at least one of the keys K1 or K2 is an incorrect key that will be subsequently eliminated from the search space.

The Double-DIP attack makes use of a more sophisticated double-miter circuit to compute 2-DIPs. As illustrated in Fig. 6.3b, the double-miter comprises two basic miter circuits and computes four key values. The output $diff = 1$ when $O_A \neq O_B$, $K_A \neq K_C$ and $K_B \neq K_D$. Moreover, the construction of the double-miter ensures that $O_A = O_C$ and $O_B = O_D$; note that the last two constraints can be incorporated in the CNF formula without any additional clauses [4]. Once a 2-DIP I_d has been computed by the SAT solver, it is applied to the functional IC and the correct output O_d is recorded. Under the specified constraints, at most two key values can produce

Fig. 6.3 (a) The miter employed by the SAT attack [5], and (b) the double-miter employed by the Double-DIP attack [4]

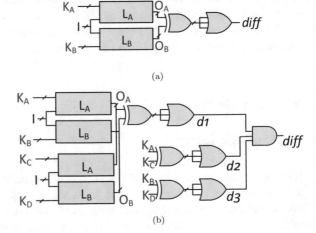

(a)

(b)

an output matching O, implying that at least two key values must lead to incorrect outputs. When the I/O pair (I_d, O_d) is added to the SAT formula, at least two key values are eliminated from the search space. A question that arises is why does Double-DIP restrict to utilizing only 2-DIPs and not use 3-DIPs, 4-DIPs or n-DIPs? In theory, it is possible to compute n-DIPs but the associated computational effort grows significantly with n. Moreover, finding the optimal value of n is still an open research question.

6.3.3   Attack Algorithm

As explained in the previous section, the Double-DIP attack uses a double-miter circuit to compute stronger *2-DIPs*, where each 2-DIP helps eliminate at least two incorrect keys in each iteration of the attack. The attack keeps iteratively applying 2-DIPs and terminates when it cannot find any further 2-DIPs; the termination of the attack guarantees that the compound logic locking technique has by then been reduced to the low-corruptibility technique. Since only 1-DIPs remain as the feasible source of injected error, the output of the recovered approximate netlist will differ from that of the original netlist for only a handful of input patterns for any key value. The flowchart of the attack is presented in Fig. 6.4. The Double-DIP attack may run into scalability issues if the correct key value for the locked circuit is not unique (refer to [4] for further details).

Fig. 6.4 Flowchart of the Double-DIP attack [4]

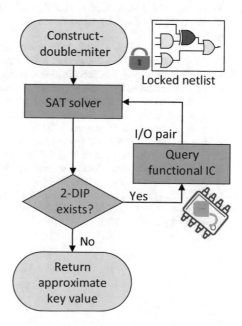

Example Let us consider the application of the Double-DIP attack to the circuit in Fig. 6.1, whose output is presented in Table 6.1. Let us assume that in the first iteration, the SAT solver returns the pattern 000 as a 2-DIP. This pattern is a valid 2-DIP as it will eliminate more than two incorrect key values. When the I/O pair $(000, 0)$ is added as a constraint to the SAT solver, the only key values that remain feasible are $k8 - k11$. However, none of the remaining input patterns can eliminate two or more keys. As an example, consider the pattern 010; the pattern is a 1-DIP since it can eliminate only $k9$. Thus, the attack terminates, returning one (e.g., $k9$) out of the feasible key values as an approximate key.

To summarize, the approximate attacks on logic locking target naive integration of low output corruptibility techniques with high output corruptibility techniques. Note that these attacks cannot break the point function-based logic locking techniques in the exact sense. The effectiveness of these attacks against compound logic locking suggests deriving high output corruptibility as an inherent property of a logic locking technique.

References

1. Rajendran J, Zhang H, Zhang C, Rose G, Pino Y, Sinanoglu O, Karri R (2015) Fault analysis-based logic encryption. IEEE Trans Comput 64(2):410–424
2. Roy J, Koushanfar F, Markov IL (2010) Ending piracy of integrated circuits. IEEE Comput 43(10):30–38
3. Shamsi K, Li M, Meade T, Zhao Z, Z D, Jin Y (2017) AppSAT: approximately deobfuscating integrated circuits. In: IEEE international symposium on hardware oriented security and trust, pp 95–100
4. Shen Y, Zhou H (2017) Double DIP: re-evaluating security of logic encryption algorithms. Cryptology ePrint Archive, Report 2017/290, http://eprint.iacr.org/2017/290
5. Subramanyan P, Ray S, Malik S (2015) Evaluating the security of logic encryption algorithms. In: IEEE international symposium on hardware oriented security and trust, pp 137–143
6. Xie Y, Srivastava A (2016) Mitigating SAT attack on logic locking. In: International conference on cryptographic hardware and embedded systems, pp 127–146
7. Yasin M, Mazumdar B, Rajendran J, Sinanoglu O (2016) SARLock: SAT attack resistant logic locking. In: IEEE international symposium on hardware oriented security and trust, pp 236–241
8. Yasin M, Sengupta A, Nabeel MT, Ashraf M, Rajendran J, Sinanoglu O (2017) Provably-secure logic locking: from theory to practice. In: ACM/SIGSAC conference on computer & communications security, pp 1601–1618
9. Zaman M, Sengupta A, Liu D, Sinanoglu O, Makris Y, Rajendran JJV (2018) Towards provably-secure performance locking. In: IEEE design, automation & test in Europe conference & exhibition, pp 1592–1597

Chapter 7
Structural Attacks

Abstract This chapter is about structural attacks on point function-based logic locking. These attacks rely on the structural properties of a locked netlist to identify the correct functionality of the original version. The chapter presents four attacks: the signal probability skew (SPS) attack, the AppSAT-guided removal (AGR) attack, sensitization-guided SAT (SGS) attack, and the Bypass attack. The SPS attack targets the basic Anti-SAT block; the AGR attack circumvents the functional and structural obfuscation added on top of basic Anti-SAT; the SGS attack exposes the security vulnerabilities associated with AND-tree detection (ATD); the Bypass attack integrates the Double-DIP attack with simple post-processing steps to recover an exact netlist.

Structural/removal attacks on point function-based logic locking exploit the structural traces embedded in a netlist to identify and/or bypass the protection offered by the point-function and recover the correct functionality of the target netlist. This chapter describes the operation of four structural attacks on logic locking. Section 7.1 presents the SPS attack that can identify and remove the basic (unobfuscated) Anti-SAT block to retrieve the original circuit. Section 7.2 elaborates on the operation of the AGR attack that integrates AppSAT with simple netlist analysis to break obfuscated Anti-SAT (OA). Section 7.3 presents the SGS attack that exploits the security vulnerabilities of ATD to weaken the security it promises. Section 7.4 discusses how the Bypass attack recovers an exact netlist by adding a bypass circuit to an approximate netlist.

7.1 Signal Probability Skew (SPS) Attack

7.1.1 Basic Idea

The SPS attack breaks the basic Anti-SAT [10] by leveraging structural traces in a netlist to identify and remove the Anti-SAT block within minutes [12]. The attack uses the notion of signal probability to identify the output gate of the Anti-SAT.

© Springer Nature Switzerland AG 2020
M. Yasin et al., *Trustworthy Hardware Design: Combinational Logic Locking Techniques*, Analog Circuits and Signal Processing,
https://doi.org/10.1007/978-3-030-15334-2_7

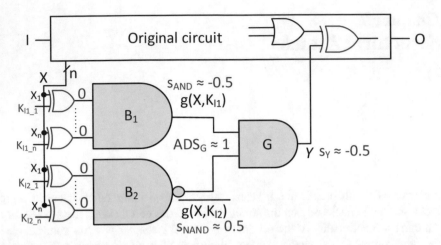

Fig. 7.1 The basic Anti-SAT block comprising two complementary functions [10]. The AND-tree and the NAND-tree are skewed in opposite directions as indicated by SPS values of -0.5 and 0.5, respectively

As already described in Sect. 5.3 and elaborated in Fig. 7.1, the Anti-SAT block comprises an AND-tree and a NAND-tree. The output of the AND-tree is highly skewed towards 0, and that of the NAND-tree is highly skewed towards 1. Thus, the output gate of the Anti-SAT block has the property that its inputs are highly skewed in opposite directions. This property is rare in realistic circuits and helps distinguish the Anti-SAT logic from the rest of the circuit.

Threat Model The threat model of the SPS attack is weaker than that of the SAT attack [9] and Anti-SAT [10]. The attack requires only a reverse-engineered netlist; it does not require access to a functional IC.

7.1.2 Preliminaries: Signal Probability Skew

We can define *signal probability skew* s_x of a signal x as,

$$s_x = Pr[x = 1] - 0.5 \tag{7.1}$$

where, $Pr[x = 1]$ denotes the probability of the signal x being 1. As $0 \leq Pr[x = 1] \leq 1$, the range of s is $[-0.5, 0.5]$. The SPS of a signal denotes the amount by which a signal is distinguishable from a random guess, which is characterized by $Pr[x = 1] = 0.5$. An attacker has a negligible advantage of guessing the signal value over a random guess if s_x is close to zero. For instance, all primary inputs and key inputs (unknown to the attacker) are equiprobable; hence, their skew is zero.

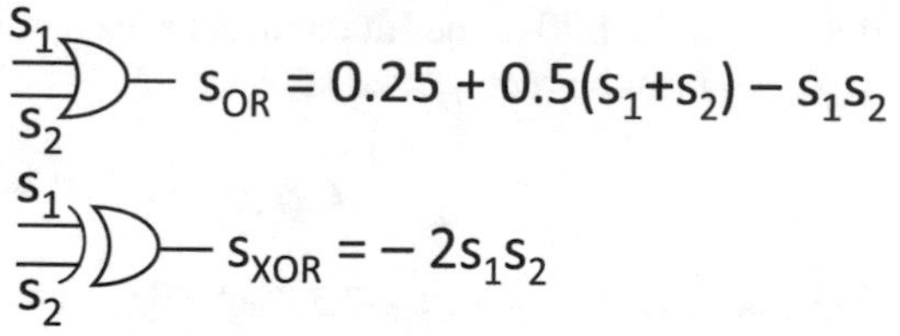

Fig. 7.2 SPS of OR and XOR gates, where s_1 and s_2 are the SPS values at the inputs of the gates

$$s_{OR} = 0.25 + 0.5(s_1 + s_2) - s_1 s_2$$

$$s_{XOR} = -2s_1 s_2$$

Consider a 2-input AND gate with inputs in_1 and in_2, and the corresponding SPS values denoted as s_1 and s_2, respectively. The SPS of the output s_{AND} is defined as:

$$s_{AND} = Pr[y = 1] - 0.5 = Pr[in_1 = 1]Pr[in_2 = 1] - 0.5$$
$$= 0.5(s_1 + s_2) + s_1 s_2 - 0.25 \tag{7.2}$$

With both inputs to an AND gate exhibiting zero skew values, $s_{AND} = -0.25$, representing the skew that an AND gate introduces into the circuit. The SPS values of OR and XOR gates are depicted in Fig. 7.2. It can be noted that an OR gate introduces a positive skew, while an XOR gate reduces the absolute skew, restoring it closer to zero. XOR/XNOR key gates, where the key inputs are treated as primary inputs, introduce a skew of zero. In MUX-based logic locking [6], the select input of a MUX is a key input with zero skew; the data inputs are internal signals in the original circuit. With s_1 and s_2 denoting the SPS at the MUX inputs, the SPS of a MUX output can be derived as,

$$s_{MUX} = 0.5(s_1 + s_2) \tag{7.3}$$

7.1.3 Attack Algorithm

In this section, we present the SPS attack algorithm that detects the output signal Y of the Anti-SAT block. We show that the *absolute difference of the probability skew (ADS)* of the inputs of a gate is the maximum for the gate G, which produces the output Y of the Anti-SAT block.

Let us consider the SPS values of the individual gates in the Anti-SAT block shown in Fig. 7.1. The XOR key gates produce zero skew signals. The blocks $g(X, K_{l1})$ and $\overline{g(X, K_{l2})}$ comprise an n-input AND and an n-input NAND gate, respectively. The SPS s_{n-AND} for an n-input AND gate is defined as,

$$s_{n-AND} = \prod_{i=1}^{n}(0.5 + s_i) - 0.5 \tag{7.4}$$

where s_i is the SPS of the ith input. As $s_i = 0$, the SPS of the AND gate constituting $g(X, K_{l1})$ can be represented as,

$$s_{g(X,K_{l1})} = 0.5^n - 0.5 \tag{7.5}$$

For a large n, $s_{g(X,K_{l1})} \approx -0.5$, indicating $p \approx 1$, where p denotes the cardinality of the onset of the function g. Similarly, for an n-input NAND gate, the SPS is,

$$s_{n-NAND} = 0.5 - \prod_{i=1}^{n}(0.5 + s_i) \tag{7.6}$$

As $s_i = 0$, the SPS of the NAND gate constituting $\overline{g(X, K_{l1})}$ is,

$$s_{\overline{g(X,K_{l1})}} = 0.5 - 0.5^n. \tag{7.7}$$

For a large n, $s_{\overline{g(X,K_{l1})}} \approx 0.5$, indicating $p \approx 2^n - 1$. The *absolute difference of the probability skew at the inputs* of the AND gate G can be represented as:

$$= |s_{g(X,K_{l1})} - s_{\overline{g(X,K_{l1})}}| = 1 - 2 \times 0.5^n \tag{7.8}$$

For a large n, $ADS_G = |s_{g(X,K_{l1})} - s_{\overline{g(X,K_{l1})}}| \approx 1$. ADS_G close to 1 indicates that the two inputs of the gate G exhibit the highest skews but with opposite polarity. This property of the gate G distinguishes it from the rest of the gates, not only in the Anti-SAT block but also in the entire circuit. The SPS attack on a circuit with the Anti-SAT block requires the computation of the ADS_G values of all the gates in the circuit. *The gate with the highest ADS_G value, i.e., the gate with the most oppositely skewed inputs, is the candidate gate G that constitutes the output gate of the Anti-SAT block.* The SPS attack is described in Algorithm 5.

Filtering the Candidate Gates In SPS attack, the gate G is identified using the highest ADS value. The original circuit may contain a few signals that exhibit high ADS values, close to ADS_G. These false candidates can be filtered out by checking

---

Algorithm 5: Signal probability skew attack

Input : $N_{antisat}$ // Netlist locked with basic Anti-SAT
Output: N_{rec} // Netlist recovered after removing the Anti-SAT block
1 $ADS_{arr} \leftarrow \{\}$

2 **for** $g_j \in N_{antisat}$ **do**
3 $\quad$|$\quad ADS_{arr}(g_j) \leftarrow$ compute_ADS($N_{antisat}, g_j$)
4 **end**
5 $G \leftarrow$ find_maximum (ADS_{arr}) // Anti-SAT output

6 $Y \leftarrow$ find_value_from_skew (G) // Correct value of Y

7 $N_{rec} \leftarrow$ remove_TFI($N_{antisat}, G, Y$) // Remove the gates that are in TFI of gate G alone

for simple structural properties. By analyzing the transitive fan-in of the candidate gates and eliminating those gates whose TFI does not include at least $2n$ key inputs, we can correctly identify the gate G.

Identifying the Value of the Anti-SAT Output Once the gate G has been identified, the value of the output signal Y can be determined from s_Y. If $s_Y < 0$, the value of Y in the functional IC is 0; otherwise, it is 1. Knowing the correct value of Y, one can trace back and discard the gates that are in the TFI of signal Y alone. The remaining circuit can be re-synthesized using the correct value of Y. Upon removal, the Anti-SAT stripped circuit can be represented as $O = F(I)$. If instead of the original circuit, the Anti-SAT block is integrated with a locked circuit, the Anti-SAT-stripped circuit is represented as $O = F(I, K_{l0})$, with K_{l0} denoting the FLL key. To identify K_{l0} for the logic locked circuit, SAT attack may be launched.

SPS vs. SAT Attack Resilience SPS attack is highly effective when $p \in \{1, 2^n - 1\}$; these values of p also lead to the maximum ADS_G. One option to reduce the effectiveness of the SPS attack is to use a value of p far from 1 and $2^n - 1$, reducing the signal skew values. However, as dictated by Eq. 5.1, any such attempt would make Anti-SAT vulnerable to SAT attack. Anti-SAT is thus cornered by the SAT attack and the proposed SPS attack.

7.1.4 Limitations

As described in Sect. 5.3.3, Anti-SAT utilizes additional functional and structural obfuscation to thwart simple removal attacks. The SPS attack is successful against OA as long as ADS_G values do not deviate significantly as a consequence of obfuscation. Let us consider an n-input AND gate that constitutes the function g in the Anti-SAT block. In Fig. 7.3a, the XOR key gate is inserted at a net inside the AND-tree, at the input of final AND gate in this specific case. Let us assume s_1 and s_2 denote the skew at the inputs of the final AND gate. Prior to insertion of the key gate, $s_1 = s_2 = 0.5^{\frac{n}{2}} - 0.5$, and $s_{n-AND} = 0.5^n - 0.5$ for the AND-tree. After

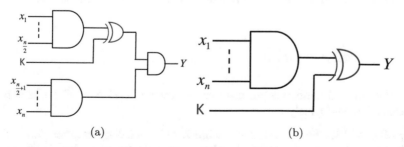

(a)   (b)

Fig. 7.3 (a) Key gate inserted inside an n-input AND-tree. (b) Key gate inserted at the output of AND-tree renders the SPS attack ineffective

the insertion of the key gate, $s_1 = 0$, and hence, the modified skew of the n-input AND becomes $s'_{n-AND} = 0.5^{\frac{n}{2}+1} - 0.5$. When the key gate is moved further to the output of AND gate as shown in Fig. 7.3b, $s_Y = 0$. The SPS attack would not be able to identify the gate G in such scenarios. Thus, by carefully inserting the key gates for functional/structural obfuscation, a designer can defend against the SPS attack. However, as illustrated in the next section, the obfuscated Anti-SAT is susceptible to a more sophisticated AGR attack.

7.2   AppSAT-Guided Removal (AGR) Attack

7.2.1   Basic Idea

The AGR attack targets OA and aims at identifying and ultimately removing the output gate of Anti-SAT, even in the presence of functional and structural obfuscation. As we discussed in Sect. 5.3.3, apart from the $2k$ XOR key gates in the basic Anti-SAT block, the OA has: (1) k additional XOR key gates for functional obfuscation, i.e., to break the symmetry of the signals in the Anti-SAT, and (2) k MUX key gates for structural obfuscation, i.e., to obscure the boundary between the Anti-SAT block and the original circuit. Moreover, OA is integrated with an FLL circuit, with the number of FLL gates being 5% of the total number of the gates in the circuit [10]. These additional key gates alter the signal probabilities in the netlist significantly, rendering the SPS attack ineffective.

The AGR attack relies on the observation that by integrating Anti-SAT with FLL, OA is essentially an instance of compound logic locking. Approximate attacks, when applied on OA, can recover only an approximate netlist [7, 8]. AGR takes this one step further and recovers an exact netlist [12]. First, it launches the AppSAT attack on the OA block, which helps distinguish the Anti-SAT key bits from the FLL key bits. Next, it applies heuristics based structural analysis, which identifies the output gate of the Anti-SAT block.

Threat Model   The threat model for the AGR attack is the same as that of the SAT attack [9]. The attacker has access to a locked netlist and a functional IC.

7.2.2   Attack Algorithm

The AGR attack algorithm comprises two main stages: (1) key bit classification and (2) transitive fan-in analysis.

Key Bit Classification   Recall that key bits in OA can belong to either FLL or Anti-SAT. AGR utilizes AppSAT to distinguish between the key bits belonging to the two classes. As the AppSAT attack proceeds, the values of the FLL key bits start being

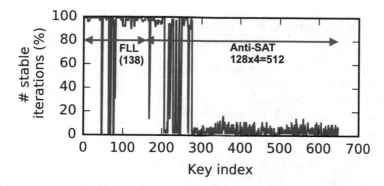

Fig. 7.4 Stability of key bits during the AppSAT attack for the c5315 FLL(5%)+128-bit OA circuit. The FLL key bits mostly remain stable and are easily distinguishable from the Anti-SAT key bits

resolved. Over successive attack iterations, the values of the FLL key bits remain stable, whereas the Anti-SAT key bits remain unresolved and their values keep fluctuating across attack iterations. Thus, key bit stability over successive attack iterations serves as a heuristic to classify the key bits into two classes.

Example Figure 7.4 demonstrates the stability of the key bits for the c5315 (5% FLL + 128-bit OA) circuit upon launching the AppSAT attack. Note that with $k = 128$, OA has a total of $4k = 512$ key bits; the number of FLL key gates is 138, resulting in a total of 650 key bits for the overall circuit. The figure displays the percentage of consecutive AppSAT iterations over which the value of a key bit (with a given index) has remained stable; as soon as a key bit value changes, the corresponding stability counter is reset to zero; otherwise, the counter is incremented in each iteration. It can be observed that most of the Anti-SAT key bits keep fluctuating and are easily distinguishable from the FLL key bits, which remain relatively stable. However, AppSAT can only partially distinguish the FLL key bits from the Anti-SAT key bits. Similar to the FLL key bits, a small number of Anti-SAT key bits (particularly those close to the AntiSAT block output) remain relatively stable over many iterations. Since the stable key bits could belong to either Anti-SAT or FLL, AGR uses only the fluctuating key bits for structural analysis.

Transitive Fan-in Analysis Having peeled off the FLL layer through AppSAT, the AGR attack identifies the output gate G of the obfuscated Anti-SAT block through simple structural analysis. The OA block has $4k$ key inputs, all of which converge at the gate G. Thus, the gate G can be identified by tracing the transitive fan-out (TFO) of the unstable Anti-SAT key inputs. One can expect around $C_g = 4k$ fluctuating key bits to converge at the gate G and around $2k$ keys bits to converge at each of its inputs, which are driven by the two complementary trees. At the inputs of gate G, each of the ratios $R_1 = \frac{C_{in1}}{C_g}$ and $R_2 = \frac{C_{in2}}{C_g}$ is close to 0.5; C_x represents the number of fluctuating keys that converge at a given gate. As depicted in Algorithm 6,

---

Algorithm 6: AppSAT guided removal attack

Input : $N_{antisat}$ // Locked netlist with Anti-SAT
Input : n // Key size for Anti-SAT
Output: N_{lock} // Locked netlist after removing Anti-SAT
1   $\#cand \leftarrow \text{num\_gates}(N_{antisat})$

2   **while** *(#cand > 1 and !timeout)* **do**
3   launch_appsat(4) // make 4 appsat calls

4   *candidates*= { }

5   **for** $g_j \in N_{antisat}$ **do**
6   **if** $N_{g_j} \approx 4n$ *and* $R_1(g_j) \approx R_2(g_j) \approx 0.5$ **then**
7   add g_j to *candidates*
8   **end**
9   **end**
10 **end**
11 $G \leftarrow$ find_maximum_key_count *(candidates)* // sort candidates by C_g and pick the top-ranking candidate

12 $N_{lock} \leftarrow$ remove_TFI($N_{antisat}$, G) // Remove the gates that are exclusively in the TFI of the gate G

---

by computing these ratios for each gate in the circuit and keeping only those close to 0.5, one can potentially identify the gate G. If this analysis returns multiple candidate gates, we can sort them based on the number of key inputs that converge at a gate and pick the first candidate as the gate G. The empirical results indicate that these simple heuristics can successfully locate the gate G in 100% of the cases [12]. Upon the identification of the gate G, an attacker can re-synthesize the locked netlist, once with a value of 0 for the gate G and once with a value of 1, retrieving a netlist that contains only the FLL key bits. Launching the SAT attack on (at least one of) these netlists will resolve the values of the FLL key bits, retrieving the original netlist.

7.3   Sensitization-Guided SAT (SGS) Attack

7.3.1   Basic Idea

As discussed in Sect. 5.4, ATD searches for AND/OR-trees inside an original netlist and locks the tree inputs using XOR/XNOR key gates [3]. When a tree of the desired size cannot be found in the circuit, a dummy AND-tree is inserted into the netlist. The SGS attack makes use of the VLSI test principle of sensitization to expose the limitations of ATD and circumvent the protection it offers, recovering the correct key values using a small number of DIPs ($\ll 2^k$), with k denoting the AND-tree size. Before describing the attack, we would highlight the security vulnerabilities of ATD.

Threat Model   The threat model for the SGS attack is the same as that of the SAT attack [9] or ATD [3]. The attacker is an end-user who has access to a locked netlist and a functional IC.

7.3.2   Security Vulnerabilities of ATD

For ATD to be effective against SAT and other attacks, the following aspects must be considered.

Existence of Large Non-decomposable Trees   As discussed in Sect. 5.4.1, the security of ATD is dictated by the size of the largest non-decomposable tree in the circuit. Large non-decomposable trees (e.g., with 64 or 128 inputs) are required to attain sufficient security against the SAT attack. Such large trees rarely exist in typical benchmark circuits. Table 7.1 shows the 15 largest AND/OR trees detected in three benchmark suites, the ISCAS benchmark circuits [2], the MCNC circuits [1], and the OpenSPARC microprocessor controllers [5]. It can be observed that, except for the circuit k2, none of the circuits has sufficiently large AND-trees to avoid the need for dummy trees. Even the large AND-tree in the circuit k2 has its own shortcomings in the form of highly biased input distribution.

Dummy AND/OR-Trees   To ensure the existence of large non-decomposable AND/OR-trees, ATD inserts large dummy trees in a circuit and integrates them with the original tree identified as part of the circuit, as illustrated in Fig. 7.5. The dummy AND-tree $T_{dummy}(I, k1)$ is integrated with the original AND-tree $T_{and}(I, k2)$ in the circuit using a camouflaged OR gate. A permanent stuck-at-0 fault is introduced at the dummy input of the OR gate by manipulating the dopant polarities [3]. With the addition of the dummy AND-tree, the output of the ATD-locked circuit can be represented as $O = F'(I) \circ T_{and}(I, k2) \circ T_{dummy}(I, k1)$, where $\circ$ denotes a generic Boolean function. However, since the inserted tree is fake and is functionally disconnected from the circuit, it remains susceptible to removal attacks.

Bias in the Input Distribution   Contrary to the externally integrated AND/OR-trees in Anti-SAT or the comparator in SARLock, the inputs of an internal AND/OR-tree may not be the primary inputs. Consequently, the input distribution of the tree will be biased. The bias in the input distribution of a k-input AND-tree implies that the tree inputs take on only a subset of 2^k possible input values. This reduction is due to the logic in the transitive fan-in of the AND-tree, i.e., the logic between the primary inputs of the circuit and the AND-tree inputs. This bias in the input distribution allows an attacker to apply only a subset of DIPs, i.e., those that bring unique values to the AND-tree inputs.

Sensitization of the Injected Error   For certain incorrect key values, an error is injected at the output of locked AND-tree. However, the injected error may not always be sensitized to a primary output; the effect of the error may be masked

Table 7.1 Size of the largest AND/OR tree in the ISCAS benchmark circuits [1], the MCNC circuits [2], the OpenSPARC microprocessor controllers [5]

Benchmark	k2	s38417	s15850	des	s38584	b18	b19	tlu_mmu	lsu_stb	s13207	ifu_ifq	c2670	lsu_excp	c3540	c1908
Tree size	104	35	28	27	27	26	26	25	24	21	20	18	18	17	14

The data is presented for 15 circuits with the largest trees

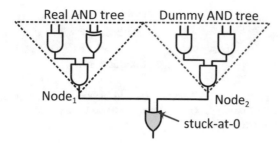

Fig. 7.5 Insertion of a dummy AND-tree into a circuit. A stuck-at-0 fault is introduced at the dummy input of the OR gate [3]

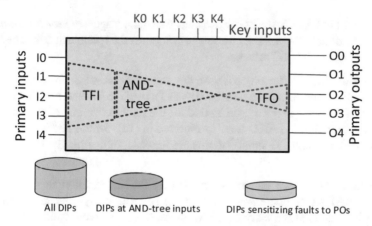

Fig. 7.6 An illustration of reduction in the number of feasible DIPs. The gates in the TFI of the tree introduce bias and reduce the number of patterns received at the tree inputs. The gates in the TFO hamper the sensitization of error injected at the output of the AND-tree and further narrow down the set of valid DIPs

by the logic in the TFO of the tree. As already discussed in the context of FLL (in Sect. 3.1), the effect of an incorrect key leading to an incorrect output is analogous to the detection of a stuck-at fault at a primary output. As described in Appendix A, the detection of a stuck-at-0 (1) fault requires that the fault be (1) activated by assigning a value 1 (0) to the fault location, and (2) propagated to a primary output.

Feasible Input Patterns The combined impact of the bias in the input distribution and the masking of an injected error is that only a subset of total 2^k DIPs is deemed *feasible*, i.e., they manifest an error in the circuit output; this is further illustrated in Fig. 7.6. Recall that the SAT attack uses the error at the output as a hint for identifying the incorrect key values [4, 9]. The smaller the fraction of input patterns used by the attack, the lower the computational effort of the attack.

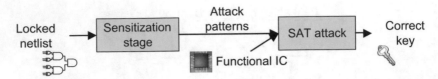

Fig. 7.7 The SGS attack on an ATD netlist. The sensitization stage computes a reduced set of attack patterns. The SAT attack uses the computed patterns in conjunction with the functional IC to determine the correct key value

7.3.3 Attack Algorithm

As illustrated in Fig. 7.7, the SGS attack has two main stages: the *sensitization* stage and the *SAT attack*. The sensitization stage computes the feasible attack patterns that are used to guide the SAT attack.

Sensitization Stage The objective of the sensitization stage is to compute the feasible input patterns that are later used as DIPs by the SAT attack. Either an ATPG tool or a SAT solver may be used to find all the patterns that can detect a stuck-at fault at the output of the AND-tree. As reported in [12], for the 104-input AND-tree in the circuit k2, only 273 input patterns are feasible, indicating the ineffectiveness of ATD.

SAT Attack The attack patterns computed by the sensitization stage are used to guide the SAT attack that extracts the correct key efficiently. The set of computed patterns is sufficient for a successful SAT attack since the set contains all the patterns that introduce observable error(s) in the circuit. The SAT attack does not need to compute further DIPs and completes within a single iteration.

Note that the SAT solvers can inherently leverage the input bias and, apparently, render the sensitization stage redundant. However, as explained next, the sensitization stage helps identify real/dummy AND-trees and prevents the SAT attack from running into long trails.

Identifying Dummy AND/OR-Trees The SGS attack tackles the challenge of identifying dummy AND/OR-trees by following a simple divide-and-conquer strategy. The attack assumes that:

1. The attacker knows the location of the key gates.
2. The dummy AND-tree inputs are the primary inputs of the circuit (or wires close to the primary inputs) since this helps minimize the bias in the input distribution [3].
3. The dummy AND-tree is large, i.e., it has 64 or more inputs.
4. None of the gates inside the dummy tree fan out to the gates in the original circuit. Only the output of AND-tree is connected to a dummy OR gate; one input of the dummy OR gate is stuck-at-0 [3], as illustrated in Fig. 7.5.

The SGS attack uses the notion of input bias to distinguish the dummy trees from the real ones; the bias is represented by the number of feasible input patterns $\#DIP_{feas}$. In the sensitization stage, $\#DIP_{feas}$ for each tree is computed (e.g., using the sharpSAT solver); the tree with the larger $\#DIP_{feas}$ is assumed to be dummy and is removed from the netlist, reducing the locked netlist to $F'(I) \circ T_{and}(I, K)$, where T_{and} denotes the real AND-tree. A successful SAT attack on the resultant netlist and the retrieval of the correct key validates the choice of the dummy AND-tree.

The correctness of the retrieved key is validated using the following simple strategy: the attacker utilizes the key value to determine the input pattern for which the locked AND-tree will output a 1. The computed pattern is applied to the functional IC to verify the circuit operation. Note that the attacker needs to verify the circuit operation for only one input pattern; the tree output is 0 for the rest of the input patterns. In case the verification process fails and the assumption about a tree being dummy turns out to be incorrect, the experiment can be repeated by switching the real and dummy trees.

Note that the SAT attack may also be applied directly on a tree without pre-computation of $\#DIP_{feas}$. However, there is a chance that the SAT attack runs into long trails. Pre-computation of the feasible input patterns by the sensitization stage prevents such situations. To summarize, the SGS attack exposes the limitations of ATD. Similar to Anti-SAT and SARLock, ATD remains susceptible to removal attacks.

7.4 Bypass Attack

7.4.1 Basic Idea

The Bypass attack relies on the observation that for any given key value, the point function-based logic locking techniques inject an error in the circuit for only a handful of input patterns [11]. Upon identifying the error-injecting input patterns, an attacker can build a small "Bypass" circuit that nullifies the error introduced by the locked circuit, thus restoring the correct circuit output. This concept is further illustrated in Fig. 7.8.

Constructing a bypass circuit for "standalone" point function-based logic locking techniques is straightforward and the required logic is relatively small (e.g., an AND-tree to restore the circuit for only one input pattern in SARLock). However, building a bypass circuit for compound logic locking requires significantly higher computational effort and the required logic may be prohibitively large, dictated mainly by the number of patterns that lead to an incorrect output. To circumvent compound logic locking, the Bypass attack first launches the Double-DIP attack (which is an approximate attack) on the locked netlist and retrieves an approximate netlist. The bypass circuitry is then constructed to restore the approximate circuit output for a single DIP.

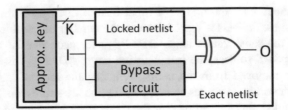

Fig. 7.8 The basic principle of the Bypass attack. An incorrect (approximate) key is supplied to the locked circuit, producing an incorrect output for specific input patterns. A Bypass circuit is constructed to restore the output for those patterns

7.4.2   Attack Algorithm

The Bypass attack consists of the following three steps: (1) launching the Double-DIP attack, (2) generating the error-injecting patterns, and (3) synthesizing the bypass circuit.

Double-DIP Attack As described in detail in Sect. 6.3, the Double-DIP attack is an approximate attack that reduces a compound logic locking technique to the constituent point function-based locking technique [8]. The key returned by the this stage results in an approximate netlist. The approximate netlist is transformed to an "exact" netlist in the subsequent attack steps.

Generating Error-Injecting Patterns In this step, the attacker takes the approximate netlist, chooses an incorrect key value as the unlock key and finds the patterns that inject error for the selected key value. There is an intermediate step involved in this computation that utilizes the miter circuit. The attacker selects another key value randomly and uses the SAT solver to compute all input patterns that lead to a differing output for the two key values. The attacker then applies the computed patterns to the functional IC and records the correct output. The patterns for which the selected unlock key generates an incorrect output are marked as error-injecting patterns.

Example Let us reconsider Table 7.2, which is a reproduction of Table 5.2 and represents the outputs of a netlist locked with SARLock. With SARLock being a point function-based technique, the Double-DIP attack will complete in a single iteration. Let's assume the attack returns $k0$ as an approximate key. To identify the error-injecting patterns for $k0$, an attacker chooses $k0$ and $k1$ (highlighted in gray) as constraints to the miter circuit and determines that 000 and 001 as the differing patterns for the selected key values. He/she then applies 000 and 001 to the functional IC and determines that $k0$ injects an error only for the pattern 000.

Synthesizing the Bypass Circuit The bypass circuit must correct the incorrect circuit output for the error-injecting pattern(s). When the correction is required for only one input pattern, the bypass circuity could be a simple AND gate which

Table 7.2 SARLock resilience against the SAT attack

abc	O	k0	k1	k2	k3	k4	k5	k6	k7
000	0	1	0	0	0	0	0	0	0
001	0	0	1	0	0	0	0	0	0
010	0	0	0	1	0	0	0	0	0
011	1	1	1	1	0	1	1	1	1
100	0	0	0	0	0	1	0	0	0
101	1	1	1	1	1	1	0	1	1
110	1	1	1	1	1	1	1	1	1
111	1	1	1	1	1	1	1	1	0

All incorrect entries occur across the diagonal. The output is always correct for the correct key value $k6$

outputs a 1 for a the specified pattern; the output of the bypass circuit is XORed with that of the locked netlist, resulting in an exact netlist.

Example For the example in Table 7.2, 000 is the only error-injecting pattern with $k0$ being the unlock key. To restore the injected error, the Bypass circuit can be constructed using a 3-input NOR gate, which outputs a 1 only for the input pattern 000.

As the number of error-injecting patterns increases, there is a linear increase in the size of the bypass circuit. While the overhead is relatively small when targeting SARLock or basic Anti-SAT for small values of p, it can grow prohibitively larger for obfuscated Anti-SAT, where the number of error-injecting patterns could be 50,000 or higher [11].

In summary, the removal attacks on logic locking can circumvent all point function-based logic locking techniques by exploiting the structural properties of the locked netlist. Thus, leaving the original circuit as is on-chip and implementing logic locking as a wrapper around the original circuit is not advisable. In Chap. 9, we describe SFLL that offers an inherent protection against the removal attacks.

References

1. Brglez F, Bryan D, Kozminski K (1989) Combinational profiles of sequential benchmark circuits. In: IEEE international symposium on circuits and systems, pp 1929–1934
2. Hansen MC, Yalcin H, Hayes JP (1999) Unveiling the ISCAS-85 benchmarks: a case study in reverse engineering. IEEE Design Test Comput 16(3):72–80
3. Li M, Shamsi K, Meade T, Zhao Z, Yu B, Jin Y, Pan D (2016) Provably secure camouflaging strategy for IC protection. In: IEEE/ACM international conference on computer-aided design, pp 28:1–28:8
4. Massad M, Garg S, Tripunitara M (2015) Integrated circuit (IC) decamouflaging: reverse engineering camouflaged ICs within minutes. In: Network and distributed system security symposium
5. Oracle (2015) OpenSPARC T1 Processor. http://www.oracle.com/technetwork/systems/opensparc/opensparc-t1-page-1444609.html [Nov 1, 2015]

 6. Rajendran J, Zhang H, Zhang C, Rose G, Pino Y, Sinanoglu O, Karri R (2015) Fault analysis-based logic encryption. IEEE Trans Comput 64(2):410–424
 7. Shamsi K, Li M, Meade T, Zhao Z, Pan DZ, Jin Y (2017) AppSAT: approximately deobfuscating integrated circuits. In: IEEE international symposium on hardware oriented security and trust, pp 95–100
 8. Shen Y, Zhou H (2017) Double DIP: re-evaluating security of logic encryption algorithms. Cryptology ePrint Archive, Report 2017/290. http://eprint.iacr.org/2017/290
 9. Subramanyan P, Ray S, Malik S (2015) Evaluating the security of logic encryption algorithms. In: IEEE international symposium on hardware oriented security and trust, pp 137–143
10. Xie Y, Srivastava A (2016) Mitigating SAT attack on logic locking. In: International conference on cryptographic hardware and embedded systems, pp 127–146
11. Xu X, Shakya B, Tehranipoor M, Forte D (2017) Novel bypass attack and BDD-based tradeoff analysis against all known logic locking attacks. In: International conference on cryptographic hardware and embedded systems, pp 189–210
12. Yasin M, Mazumdar B, Sinanoglu O, Rajendran J (2017) Removal attacks on logic locking and camouflaging techniques. IEEE Trans Emerg Top Comput 99(0):PP

Chapter 8
Post-SAT 2: Insertion
of SAT-Unresolvable Structures

Abstract This chapter presents cyclic logic locking and one-way function-based logic locking. The underlying idea of both schemes is to embed structures in a netlist that are hard to resolve for a SAT solver. Cyclic logic locking introduces cycles into a netlist with the expectation that it will render the SAT attack effort exponential in the number of cycles introduced. However, cyclic logic locking is vulnerable to the CycSAT attack, which can encode the presence of cycles in the CNF representation. One-way function-based logic locking integrates one-way functions into the locked netlist to render the SAT attack computationally infeasible.

The SAT attack resilient techniques discussed so far achieve high SAT attack resilience by compromising on the output corruptibility. This chapter introduces two logic locking techniques that need not make such compromise. Section 8.1 introduces cyclic logic locking that inserts cycles/loops in a netlist to thwart the SAT attack. Section 8.2 highlights the security vulnerabilities of cyclic logic locking and presents CycSAT, an attack that can break cyclic logic locking. Section 8.3 presents one-way function-based logic locking that integrates one-way functions into the locked netlist.

8.1 Cyclic Logic Locking

8.1.1 Basic Idea

The assumption underlying cyclic logic locking is that cycles are hard to encode using the CNF representation and thus unresolvable for a SAT solver. By introducing cycles in a combinational circuit, the circuit can no longer be represented as a directed acyclic graph, rendering the conventional SAT attack ineffective [4]. A cyclic circuit is represented better as a flow graph. In a combinational circuit, the flow is from sources (primary inputs) towards sinks (primary outputs), i.e., all edges

© Springer Nature Switzerland AG 2020
M. Yasin et al., *Trustworthy Hardware Design: Combinational Logic
Locking Techniques*, Analog Circuits and Signal Processing,
https://doi.org/10.1007/978-3-030-15334-2_8

point forward, from inputs towards outputs. Cycles can be introduced in a graph by introducing backward edges that oppose the normal flow. However, the additional edges must not be easily identifiable or removable. Accordingly, cyclic logic locking introduces "non-reducible" cycles in a circuit such that an attacker cannot identify and remove the additional edges from the graph [4].

8.1.2   Non-reducible Cycles

The objective of cyclic logic locking is to introduce non-reducible cycles in a graph [4]. A flow-graph with only reducible cycles has the property that its depth-first-search (DFS) traversal is unique. A DFS traversal of such a graph can create two sets of the edges: the forward edges and the backward (or, in short, back) edges. The graph constructed using only the forward edges is an acyclic graph.

Example For the graph in Fig. 8.1a, there is only one entry-point to the cycle, the node g_1. A forward traversal of the graph through the entry point g_1 can traverse nodes in the order: $g_1 \rightarrow g_2 \rightarrow g_3$, identifying the edge e_3 as a back edge since the sink for this edge is the entry point of the cycle and the source is a node inside the cycle. For the graph in Fig. 8.1b, however, there are two entry points to the cycle, g_1 and g_2. Entering the cycle through g_1 renders e_3 as a back edge, whereas, entering through g_2 renders e_1 as a back edge. Thus, the graph cannot be reduced to a unique set of forward and backward edges. When an attacker removes an edge from an actual circuit, it might lead to gates without any input or output, hinting that the edge is *non-removable*; otherwise, the edge is considered removable. For example, removing the edge e_1 leaves the gate g_1 without an output, rendering the edge non-removable. From a security standpoint, each cycle must contain as many removable edges as possible. As pointed out in [4], in a cycle with n edges, m edges might be removed to "open" the cycle and make the graph acyclic; the number of ways to remove the edges is $2^n - 1$ [4].

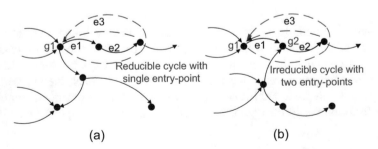

Fig. 8.1 (a) A reducible graph with g_1 being the only entry point to the cycle. (b) An irreducible graph with the cycle having two possible entry points, g_1 and g_2. The DFS traversal of the irreducible graph in not unique [4]

8.1.3 Cyclic Logic Locking Algorithm

The cyclic logic locking algorithm inserts key-controlled gates and removable dummy edges into a circuit. Upon application of incorrect key values, the dummy edges become part of the functional paths and introduce errors in the circuit behavior. For the correct key value, however, these edges do not interfere with the functional paths and the correct circuit operation is ensured. As depicted in Fig. 8.2, the dummy edges may be introduced using either AND/OR gates or muxes [4]. For the mux in Fig. 8.2b, the correct key value connects the correct wire w_0 to the circuit whereas an incorrect key value introduces a cycle in the circuit. It must, however, be ensured that the all the newly introduced edges are removable. As illustrated in Algorithm 7, cyclic logic locking aims at maximizing the number of removable edges in a cycle by inserting key-controlled muxes for all the gates in a cycle.

Example Let us consider the application of the cyclic logic locking algorithm to the original circuit in Fig. 8.3a. For the locked netlist shown in Fig. 8.3b, it is assumed that a path of length three exists from the node g_3 to the node g_1. In order to feed

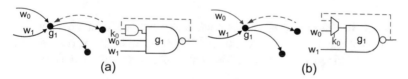

(a)

(b)

Fig. 8.2 Introducing key-controlled dummy cycles in a circuit using: (**a**) AND/OR key gates, and (**b**) Mux key gates [4]

Algorithm 7: The cyclic logic locking algorithm [4]

Input : N_{orig} // The original netlist
Output: N_{nm} // Locked netlist having n cycles, each with m
       removable edges

1 $N_{nm} \leftarrow N_{orig}$
2 **while** *cycles formed* $< n$ **do**
3 | **repeat**
4 | | $u \leftarrow$ select_random_gate(N_{nm})
5 | | Initiate DFS at u and find a path (u, v) of length m
6 | **until** *a path of length m is found*
7 | feed v back to u using a mux
8 | **foreach** *edge g on the path (u, v)* **do**
9 | | **if** $|fanout(source(g))| == 1$ **then**
10 | | | make g removable using two muxes
11 | | **end**
12 | | **else**
13 | | | make g removable using one mux
14 | | **end**
15 | **end**
16 **end**

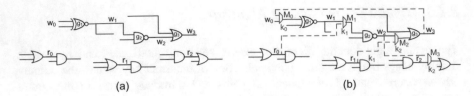

Fig. 8.3 An illustration of cyclic logic locking [4]. (**a**) The original netlist, and (**b**) the locked netlist

the signal w_3 to w_0, Algorithm 7 introduces a single mux M_0. For the net w_1 with a fan-out of 2, only one mux M_1 is sufficient to introduce a removable edge. For the net w_2 with a fan-out of 1, however, two muxes, M_2 and M_3, are utilized. Based upon the key value k_2, $M2$ selects between w_1 and a randomly selected functional wire r_1. M_3 introduces a new dummy path from w_2 to another random net.

8.1.4   Security Analysis

SAT Attack Resilience While cyclic logic locking claims strong resilience against the basic SAT attack, the CycSAT attack [7] (presented in the next section) breaks cyclic logic locking by appending additional constraints to the SAT formula.

Sensitization Attack Resilience Cyclic logic just ensures creation of cycles in a locked netlist. The interdependence among the key gates/muxes in a single cycle may hamper sensitization of certain key bits. However, the technique does not offer any guarantees in this regard and a subset of key bits might be susceptible to the sensitization attack.

Approximate Attack Resilience Cyclic logic locking does not compromise on output corruptibility to attain high SAT attack resilience and can offer high output corruptibility. However, similar to the basic logic locking techniques, such as RLL and FLL, it remains vulnerable to the approximate attacks that can easily recover an approximate netlist.

Removal Attack Resilience Cyclic logic locking distributes the protection logic all across the netlist. Removing the inserted muxes will result in an erroneous netlist, hinting at the resilience of cyclic logic locking against the removal attacks.

8.2 CycSAT

8.2.1 Basic Idea

The CycSAT attack targets cyclic logic locking and aims to retrieve a key value that renders the locked circuit acyclic and functionally equivalent to the original circuit [7]. The attack is based on the observation that the SAT attack can be executed even on cyclic circuits by adding additional constraints to the SAT formula such that the recovered key k_{rec} obtains a cycle-free circuit. Recall that the cyclic logic locking algorithm ensures, by definition, that upon application of the correct key, the locked circuit will be devoid of any cycles. Formulating these additional "No Cycle (NC)" constraints is a recursive process, as explained next.

8.2.2 Formulating NC Constraints

The computation of the NC constraints is based on determining a set of feedback signals that, when removed from the netlist, renders the circuit acyclic [7]. The feedback signals can be denoted as $s_1, s_1', s_2, s_2', \cdots, s_m, s_m'$, where s_i and s_i' represent the signals before and after the cycles have been eliminated. The overall NC constraints are a conjunction of the constraints for the individual signals.

$$NC = \bigwedge_{i=1}^{m} NF(s_i, s_i') \tag{8.1}$$

where $NF(u, v)$ is a recursive constraint depicting that there exists "no feedback" path from the signal u to the signal v. This recursive constraint must ensure that there is no feedback from u to v through any of the non-key fan-ins of v. Furthermore, for each key-gate on a path from u to v, a key constraint must be introduced such that v is not affected by any of its fan-in signals, essentially removing v from the circuit.

$$NF(u, v) = \bigwedge_{w \in NKF(v)} NF(u, w) \vee KC(w, v) \tag{8.2}$$

Here, $NKF(w)$ denotes the non-key fan-ins of the signal v, and $KC(w, v)$ is key value constraint to ensure that w does not affect v. Note that,

$$NF(u, u) = 0 \tag{8.3}$$

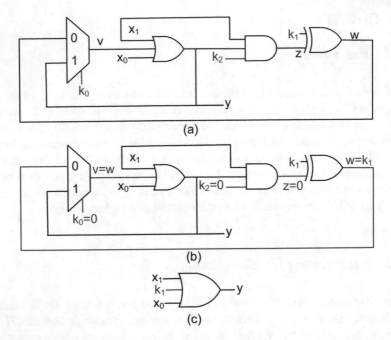

Fig. 8.4 Deriving NC constraints for a circuit locked using cyclic logic locking [7]. (**a**) A circuit locked with cyclic logic locking. (**b**) Applying the derived key constraints to the locked circuit. (**c**) The simplified acyclic circuit

Example Let us apply the aforementioned constraint formulation to the circuit in Fig. 8.4. There are two possible sets of feedback signals: $\{v\}$, and $\{w, y\}$. Applying key constraints that exclude either of these sets from the circuit results in an acyclic circuit. Focusing on $\{v\}$, we can formulate the following the NF constraints as follows.

$$NF(v, v) = 0$$
$$NF(v, y) = NF(v, v) \vee KC(v, y) = 0$$
$$NF(v, z) = (NF(v, y) \vee KC(y, z)) \wedge (NF(v, x_1) \vee KC(x_1, z)) = !k_2$$
$$NF(v, w) = NF(v, z) \vee KC(z, w) = !k_2 \tag{8.4}$$
$$NF(v, v') = (NF(v, w) \vee KC(w, v')) \wedge (NF(v, y) \vee KC(y, v'))$$
$$= (!k_2 \vee k_0) \wedge (!k_0)$$

Since $\{v\}$ is a singleton set, $NC = NF(v, v')$.

$$NC = NF(v, v') = \quad (!k_2 \vee k_0) \wedge !k_0 \tag{8.5}$$
$$= \quad !k_2 \wedge !k_0$$

Algorithm 8: CycSAT attack algorithm [7]

Input : Locked netlist $L(I, K)$, Functional IC $F(I)$
Output: Correct key K_{rec} that renders L acyclic
1 Determine the set of feedback signals $(s_0, s_1, \cdots, s_m)$
2 Compute NF formulas, $NF(s_0, s_0'), \cdots, NF(s_m, s_m')$
3 $NC = \bigwedge_{i=0}^{m} NF(s_i, s_i')$
4 $L(I, K_A) = L(I, K_A) \wedge NC(K_A)$
5 $L(I, K_B) = L(I, K_B) \wedge NC(K_B)$
6 **while** $I_d = SAT(L(I, K_A) \neq L(I, K_B))$ **do**
7      $O_d = F(I_d)$ // Query the oracle
8      $L(I, K_A) = L(I, K_A) \wedge (L(I_d, K_A) = O_d)$ // Augment clauses
9      $L(I, K_B) = L(I, K_B) \wedge (L(I_d, K_B) = O_d)$
10 **end**
11 $K_{rec} = SAT(L(I, K_A))$

When the constraint $(!k2 \wedge !k0)$ is applied to the locked circuit, the signal y is excluded from the circuit, leading to elimination of the mux from the circuit and simplification of the circuit as $k_1 \vee x_0 \vee x_1$ [7], as illustrated in Fig. 8.4.

8.2.3 Attack Algorithm

The CycSAT algorithm is the essentially the same as the SAT attack (depicted in Algorithm 4) except for the additional NC constraints, which ensure that there exist no cycles in the locked netlist upon application of the retrieved key k_{rec}. The CycSAT attack algorithm, depicted in Algorithm 8, challenges the basic assumption underlying cyclic logic locking technique and circumvents it by formulating the additional constraints on the locked netlist [7].

8.3 ORF-Lock: One-Way Function-Based Logic Locking

8.3.1 Basic Idea

As stated in Eq. 4.2, the SAT attack execution time is given as $T = \sum_{i=1}^{\lambda} t_i$. One of the options to maximize the computational effort the SAT attack is to utilize hard SAT instances that can significantly increase t_i, the time taken by one attack iteration. ORF-Lock simply integrates one-way random functions (ORFs) with a locked netlist. ORFs, such as hash functions or AES with a secret key, make it computationally infeasible for an attacker to determine the ORF inputs from its output [1, 3]. In terms of the SAT attack resilience, the time t_i required for the ith iteration will grow enormously as the attack proceeds over successive iterations [6].

8.3.2   Methodology

The integration of a locked netlist (locked using RLL, FLL etc.) and an ORF
instance, such as an AES block with a fixed-key, is conducted as follows. The
designer applies a random key to the AES circuit (the key is known only to the
designer) and synthesizes the fixed-key AES circuit. He/she then selects $K1$ inputs
of AES on a random basis and treats them as key inputs for the overall circuit;
the remaining AES inputs are tied to constant values (known only to the designer).
$K1$ outputs of the ORF block are then connected to selected key inputs in the locked
circuit. As the designer knows the fixed secret key to the AES as well as the constant
value applied to AES inputs, he/she can configure the XOR/XNOR key gates to
produce a correct output upon application of the secret key. As illustrated in Fig. 8.5,
the original netlist is locked with $K = K1 + K2$ key bits. Only $K1$ key inputs of
the locked netlist are attached to the output of the AES circuit, while the remaining
$K2$ key inputs are connected directly to the on-chip tamper-proof memory.

8.3.3   Security Analysis

SAT Attack Resilience   It is computationally infeasible to determine the inputs of
a one-way function from its outputs. In ORF-Lock, a subset of key inputs are the
inputs to the fixed-key AES circuit. Thus, an attacker cannot determine the values
of those key inputs from the outputs of a functional IC. Figure 8.6 demonstrates the
effectiveness of ORF-Lock using the c7552 SLL circuit that also embeds a fixed-
key AES circuit. It can be observed that the execution time of the SAT attack grows
exponentially with $K1$, which denotes the number of key inputs connected to the
AES circuit.

Sensitization Attack Resilience   In ORF-Lock, the key inputs of the locked circuit
are connected to the AES outputs, which in turn, depend on the AES key inputs
connected to the tamper-proof memory. AES is characterized by its high confusion
properties, implying that each output bit is affected by multiple key bits. Thus,
it is not feasible to apply the sensitization attack to resolve attack key bits on an
individual basis.

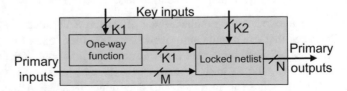

Fig. 8.5 Proposed ORF-Lock architecture that resists the SAT attack. K1 out of K key inputs in
the locked netlist are connected to the ORF circuit

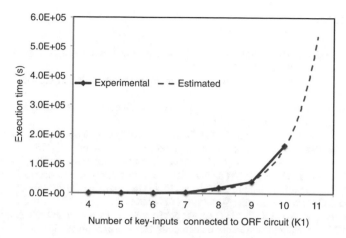

Fig. 8.6 Time required by the SAT attack [5] to break the ORF+c7552 circuits for different key inputs connected to the ORF circuit (K1). The blue line represents the best-fitting exponential curve to the experimental data

Removal Attack Resilience Since the structure of the AES is public, an attacker can remove the AES circuit, thereby extracting only the locked design. After removing the AES circuit successfully, the SAT attack [5] may be launched. However, ORF-Lock can protect against such attacks by synthesizing an AES circuit that is different from the public AES. It fixes the AES key to a randomly chosen value and synthesizes the fixed-key AES circuit (which behaves as a pseudorandom function [2]) along with the locked circuit, the functionality of which is also unknown to the attacker. Through the unified synthesis of a pseudorandom function and an unknown locked circuit, it becomes computationally infeasible to distinguish between the components on an individual basis, thwarting removal attacks.

Approximate Attack Resilience ORF-Lock does not rely on any low-output corruptibility circuits for delivering security against SAT and removal attacks and remains resilient to approximate attacks.

   In summary, this chapter presented logic locking techniques that do not compromise on output corruptibility to attain high SAT attack resilience. Cyclic logic locking relies on introducing cycles in a circuit. However, the technique is vulnerable to CycSAT. ORF-Lock uses one-way functions to thwart SAT and other attacks. However, the technique incurs large overhead that may be unacceptable in many applications. Furthermore, establishing security properties for the technique depends on the synthesis algorithm, which integrates and mixes the locked circuit with the one-way function, remains an open problem.

References

1. Goldreich O (2001) Foundations of cryptography: volume 1, basic tools. Cambridge University Press, Cambridge. ISBN 978-0521035361
2. Katz J, Lindell Y (2014) Introduction to modern cryptography. CRC Press, Boca Raton. ISBN 978-1466570269
3. Matsuzaki N, Tatebayashi M (1994) Apparatus and method for data encryption with block selection keys and data encryption keys. US Patent 5,351,299
4. Shamsi K, Li M, Meade T, Zhao Z, Pan DZ, Jin Y (2017) Cyclic obfuscation for creating sat-unresolvable circuits. In: ACM great lakes symposium on VLSI, pp 173–178
5. Subramanyan P, Ray S, Malik S (2015) Evaluating the security of logic encryption algorithms. In: IEEE international symposium on hardware oriented security and trust, pp 137–143
6. Yasin M, Rajendran J, Sinanoglu O, Karri R (2016) On improving the security of logic locking. IEEE Trans CAD Integr Circuits Syst 35(9):1411–1424
7. Zhou H, Jiang R, Kong S (2017) CycSAT: SAT-based attack on cyclic logic encryptions. In: IEEE/ACM international conference on computer-aided design, pp 49–56

Chapter 9
Post-SAT 3: Stripped-Functionality Logic Locking

Abstract This chapter presents stripped-functionality logic locking (SFLL), a technique that provides provable security against SAT, removal, and approximate attacks. SFLL hides part of the design functionality in the form of compactly represented input patterns, rendering the on-chip circuit different from the original circuit. Only upon applying the correct key(s) to the restore circuit, the original functionality of the circuit is restored.

This chapter presents SFLL, a logic locking technique that offers provable security guarantees against various classes of logic locking attacks. The underlying principle of logic locking is to implement a modified circuit on-chip; the difference in functionality is quantified in terms of the number of protected input patterns, which also dictates the protection achieved against various attacks. Only upon application of the correct key to a separately added restore circuit, the original functionality is restored. Section 9.1 explains the motivation behind SFLL and the basic concepts associated with SFLL. Section 9.2 introduces a special case of SFLL, referred to as SFLL-HD^0, which protects only one pattern. Section 9.3 elaborates on the operation of the more general SFLL-HD scheme. Section 9.4 presents SFLL-flex that allows a designer to specify the functionality-to-be-protected.

9.1 Motivation and Basic Concepts

9.1.1 Motivation

As discussed in Chaps. 5 through 7, the point function-based logic locking techniques: (1) remain susceptible to removal attacks and (2) exhibit low output corruptibility. Compound techniques that improve the output corruptibility remain vulnerable to the approximate attacks. Even cyclic logic locking can be broken using CycSAT while the overhead associated with ORF-Lock renders it impractical. Thus, there is still a need for a secure and cost-effective logic locking technique that can withstand all known and anticipated attacks. In light of the discussion in the previous chapters, a secure logic locking technique must:

© Springer Nature Switzerland AG 2020
M. Yasin et al., *Trustworthy Hardware Design: Combinational Logic
Locking Techniques*, Analog Circuits and Signal Processing,
https://doi.org/10.1007/978-3-030-15334-2_9

1. Exhibit strong resilience against the SAT attack [9].
2. Inject sufficiently large error into a circuit to achieve high output corruptibility while also thwarting approximate attacks [7, 8].
3. Circumvent removal attacks and not leave any structural traces in the circuit.
4. Incur minimal overhead in terms of power, performance, and area.

Furthermore, a logic locking framework should enable a designer to specify and protect the security-critical parts of the design IP, and thereby, to customize logic locking to the security needs of the application. It is also essential to study the connection between resilience to the SAT, removal, and approximate attacks on a quantitative basis.

To achieve the objectives mentioned above and offer holistic protection against various attacks, SFLL takes a different approach as compared to the other logic locking techniques. SFLL strips part of the design functionality from its hardware implementation. The design implemented on-chip is therefore no longer the same as the original design, as the former will be missing the stripped functionality [14]. The hardware implementation can be conceived to have a *controllable built-in error*. This error is canceled by a *restore unit* only upon the application of the secret logic locking key [14].

The stripped functionality can be captured efficiently in terms of *protected input cubes* for which the on-chip design and the original design produce differing outputs. Note that input cubes refer to partially-specified input patterns; selected input bits are specified as logic-0's or logic-1's while other input bits are don't cares (x's). An n-bit input cube with k specified (care) bits represents 2^{n-k} input patterns.

9.1.2 Variants of SFLL

Depending upon on how the input cubes are specified, SFLL has two variants: SFLL-HD and SFLL-flex [14].

- **SFLL-HD**. This variant is suitable for general applications where stripping an arbitrary part of the functionality is sufficient for protection and it is desirable to protect a large number of arbitrary input cubes. The designer specifies only one input cube (the k-bit secret key k_{sec}) and the Hamming distance h. SFLL-HD protects all input cubes that are of Hamming distance h from k_{sec} [14]. This chapter first explains SFLL-HD for the special case of $h = 0$ and later generalizes it.
- **SFLL-flex**. This variant allows a designer to specify c IP-critical input cubes to be protected, each with k specified bits. For instance, a designer may want to protect specific ranges of addresses in a processor for which access is granted only to the qualified entities [1]. SFLL-flex compresses the designer-specified input cubes and performs a security-aware logic synthesis that strips the functionality in a cost-effective manner [14].

9.2 SFLL-HD^0: A Special Case of SFLL-HD

9.2.1 Basic Idea

In this version of SFLL, there is only one protected input cube, which is the same as the secret key. SFLL-HD^0 modifies a design to invert its output for one selected (protected) input pattern; this inversion is the manifestation of the built-in error.[1] As explained in the next subsection, the restore unit cancels the built-in error to restore the correct output upon application of the correct key [14].

9.2.2 Architecture

The SFLL-HD circuit consists of a restore unit that is integrated with the functionality-stripped circuit (FSC) using an XOR gate. The restore unit computes the Hamming distance between the key inputs and the primary inputs. In the special case of SFLL-HD^0, the Hamming distance between the primary inputs and the key is zero, implying that the *restore signal* is asserted only when the key inputs and the primary inputs match [14]. The architecture of SFLL-HD^0 is depicted in Fig. 9.1. The circuit is protected by a three-bit key, $n = k = 3$; the protected cube is an input pattern, as $n = k$ in this example. The restore unit is essentially a comparator that asserts the restore signal whenever $IN = K$, i.e., across the diagonal in the truth table depicted in Fig. 9.1. The original and the functionality-stripped circuits produce a different output for only input pattern 6. The column Y_{fs} in Fig. 9.1 shows the inversion (error) for this protected input pattern. This error is cancelled out by applying the correct key k_6 which asserts the restore signal for input pattern

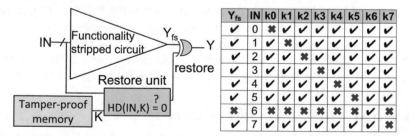

Fig. 9.1 SFLL-HD^0 architecture for $n = k = 3$. The FSC is minimally different from the original function: they produce a different response for one protected input pattern (IN = 6). The restore unit cancels this error for the correct key k6 and introduces a second error for each incorrect key. Errors are denoted by X's (red)

---

[1] SFLL-HD^0 is also referred to as TTLock, i.e., "Tenacious and Traceless Logic Locking" [15].

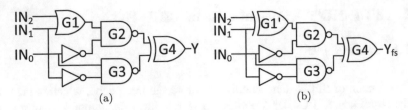

Fig. 9.2 (**a**) Original circuit, (**b**) FSC. Gate G1 is replaced with G1′ in the FSC, inverting the output Y_{fs} for IN = 6

6, and thus, recovers the desired output. Similar to the case of ATD (illustrated in Table 5.3), SFLL-HD^0 injects an error for two entries in each column, one across the diagonal and one for the protected pattern; the latter manifests as a row of incorrect entries.

The functionality-stripping can be effected via logic gate insertions or replacements, as illustrated in Fig. 9.2. In this particular example, the OR gate $G1$ has been replaced with an XOR gate $G1′$. Comparing the truth tables of a 2-input OR gates with that of a 2-input XOR gate, we find that the output of the two gates is identical except for the pattern 11, for which the OR gate produces a 1 whereas the XOR gate produces a 0. The result of this gate replacement is that the output of the original and the functionality-stripped circuit (FSC) differs for the input pattern 110; the original circuit outputs a 1, whereas the FSC outputs a 0.

9.2.3   Security Analysis

SAT Attack Resilience In each iteration of the SAT attack, a DIP (except for the protected one) eliminates exactly one incorrect key, ensuring that the required #DIPs is exponential in the key size k. In the example shown in Fig. 9.1, the attack requires $7 = 2^3 - 1$ iterations in the worst-case. However, a fortuitous attacker may try the protected input cube ($IN = 6$ in this example) and be able to eliminate all incorrect keys immediately. Since an attacker does not have any information about the protected cube, the probability of such a fortuitous hit is exponentially small in the number of key bits.

Theorem 9.1 *SFLL-HD^0 is k-secure against the SAT attack.*

Proof Let us divide the input cubes into two disjoint sets, the set of protected cubes P and the set of unprotected cubes $\widehat{P}$. For SFLL-HD^0, $|P| = 1$ *and* $|\widehat{P}| = 2^k - 1$, since there is only one protected input cube. For an attacker making only a polynomial number of queries $q(k)$ to the functional IC, the probability P_{succ} of the attack success, i.e., finding the protected input cube is given as

$$P_{succ} = \frac{|P|}{2^k} + \frac{|P|}{2^k - 1} \cdots \frac{|P|}{2^k - q(k)} \tag{9.1}$$

$$= \frac{1}{2^k} + \frac{1}{2^k - 1} \cdots \frac{1}{2^k - q(k)}$$

$$\approx \frac{q(k)}{2^k}$$

Thus, from Definition 4.1, SFLL-HD^0 is k-secure against the SAT attack.

Sensitization Attack Resilience   As already pointed out, the resilience against the sensitization attack is dictated by the number of pairwise secure key gates [5].

Theorem 9.2   *SFLL-HD^0 is k-secure against a sensitization attack.*

Proof   The proof is similar to that for SARLock (Theorem 5.2). In SFLL-HD^0, all the k bits of the key converge within the comparator inside the restore unit to produce the restore signal. Therefore, sensitizing any key bit through the restore signal to the output requires controlling all the other key bits. All k bits are therefore pairwise-secure. Thus, from Definition 4.2, SFLL-HD^0 is k-secure against sensitization attack.

Removal Attack Resilience   A removal attack launched on a locked netlist N_{lock} tries to isolate and remove the protection logic, recovering N_{rec}. A perfectly successful attack is a transformation $T : N_{lock} \rightarrow N_{rec} \mid \forall i \in I, N_{rec}(i) = F(i)$, irrespective of the key value. Note that for a partially successful removal attack, $N_{rec}(p) \neq F(p), \forall p \in P$, where P denotes the set of protected patterns. Since, SFLL is the first technique to quantify the protection against removal attacks, we introduce a formal definition of removal attack resilience here.

Definition 9.1   A logic locking technique $\mathcal{L}$ is λ-resilient against a removal attack, where λ denotes the cardinality of the set of protected patterns.

Since the restore signal in SFLL-HD^0 is highly skewed towards 0, it can be easily identified using signal probability analysis, as employed earlier by the SPS attack [13]. However, such a removal attack would recover only the FSC, without leaking any information about the original design. As the FSC produces an erroneous response for the protected input cube, the design is resilient against removal attack.

Theorem 9.3   *SFLL-HD^0 is 2^{n-k}-resilient against removal attack.*

Proof   The attacker recovers the circuit N_{rec} by identifying and removing the restore logic. N_{rec} produces an incorrect output for the set of protected input cubes P; each cube contains 2^{n-k} input patterns, leading to a total of $\Gamma = |P| \times 2^{n-k}$ input patterns such that $N_{rec}(i) \neq F(i), \forall i \in \Gamma$.

$$|\Gamma| = |P| \times 2^{n-k} \tag{9.2}$$

$$= 1 \times 2^{n-k}$$

$$= 2^{n-k}$$

Thus, SFLL-HD^0 is 2^{n-k}-resilient against a removal attack.

Note that for $n = k$, there is only one protected input pattern.

Approximate Attack Resilience Similar to ATD, FLL-HD^0 is a low-corruptibility technique since the error injected into a circuit is minimal. For any arbitrary incorrect key value, only two input patterns will generate an incorrect output, resulting in an OER of $\frac{2}{2^k}$ (assuming $n = k$).

9.3   SFLL-HD for Protecting Multiple Patterns

In this section, we generalize for h; *SFLL-HD* protects all input cubes that are of Hamming distance of h from the secret key. The number of protected input cubes is $\binom{k}{h}$. The main advantage over SFLL-HD^0 is that with increased h, the resilience to removal and approximate attacks increases significantly.

9.3.1   Architecture

With a Hamming distance h, an input size n, and a key size k, SFLL-HD inverts the FSC output Y_{fs} for $\binom{k}{h}$ input cubes, which represent $2^{n-k} \cdot \binom{k}{h}$ input patterns. The restore unit, which comprises k XOR gates and an adder to compute the Hamming distance, rectifies all these errors for the correct key, while it introduces a different but possibly overlapping set of errors for any incorrect key [14]. Figure 9.3 depicts the architecture of the proposed SFLL-HD along with an example where $n = k = 3$ and $h = 1$. As can be seen from the architecture, the hardware implementation and the associated overhead of the restore unit is independent of h, which is a hard-coded (non-secret) constant that feeds the comparator inside the restore unit.

9.3.2   Security Analysis

SAT Attack Resilience With increasing h, the number of protected patterns increases, leading to a decrease in the SAT attack resilience.

Theorem 9.4 *SFLL-HD is $(k - \lceil \log_2 \binom{k}{h} \rceil)$-secure against SAT attack.*

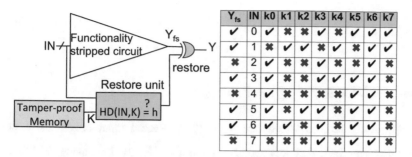

Y_{fs}	IN	k0	k1	k2	k3	k4	k5	k6	k7
✓	0	✓	✗	✗	✓	✗	✓	✓	✓
✓	1	✗	✓	✓	✗	✓	✗	✓	✓
✗	2	✓	✗	✗	✓	✗	✗	✓	✗
✓	3	✓	✗	✗	✓	✓	✓	✓	✗
✗	4	✓	✗	✗	✗	✗	✓	✓	✗
✓	5	✓	✗	✓	✓	✗	✓	✓	✗
✓	6	✓	✓	✗	✓	✗	✓	✓	✗
✗	7	✗	✗	✗	✓	✗	✓	✓	✗

Fig. 9.3 SFLL-HD architecture for $n=k=3$ and $h=1$. Y_{fs} includes $\binom{k}{h}$ errors, denoted by X's (red). Restore unit rectifies all errors for the correct key $k6$. For the incorrect keys, restore unit introduces $\binom{k}{h}$ additional errors (at $\binom{k}{h}$ input patterns), which may possibly coincide and cancel errors in Y_{fs}

Proof The proof is similar to that for Theorem 9.1. In SFLL-HD, $|P| = \binom{k}{h}$ and $|\widehat{P}| = 2^k - \binom{k}{h}$. Thus, for an attacker making only polynomial number of queries $q(k)$ to the oracle, the success probability is given as

$$P_{succ} = \frac{|P|}{2^k} + \frac{|P|}{2^k - 1} \cdots \frac{|P|}{2^k - q(k)}$$

$$= \frac{\binom{k}{h}}{2^k} + \frac{\binom{k}{h}}{2^k - 1} \cdots \frac{\binom{k}{h}}{2^k - q(k)}$$

$$\approx \frac{q(k) \cdot \binom{k}{h}}{2^k}$$

$$< \frac{q(k)}{2^{k - \lceil \log_2 \binom{k}{h} \rceil}}$$

Thus, SFLL-HD is $(k - \lceil \log_2 \binom{k}{h} \rceil)$-secure against the SAT attack.

Sensitization Attack Resilience The basic structure of restore unit for SFLL-HD is the same as that for SFLL-HD^0, implying similar resilience against the sensitization attack.

Theorem 9.5 *SFLL-HD is k-secure against sensitization attack.*

Proof The proof is identical to that for Theorem 9.2.

Removal Attack Resilience An increasing number of protected patterns implies a higher resilience against the removal attacks.

Theorem 9.6 *SFLL-HD is $2^{n-k} \cdot \binom{k}{h}$-resilient against removal attack.*

Proof Using a removal attack, an attacker recovers only the FSC, which produces an erroneous output for all the protected input patterns (Γ). Similar to Theorem 9.3, from Eq. 9.2,

$$|\Gamma| = |P| \times 2^{n-k}$$

$$= \binom{k}{h} \times 2^{n-k}$$

Thus, SFLL-HD is $2^{n-k} \cdot \binom{k}{h}$-resilient against a removal attack.

Approximate Attack Resilience With $\binom{k}{h}$ protected input cubes, the expected OER for any incorrect key can be given as $\frac{\binom{k}{h}}{2^k}$. By increasing h, OER can be increased as desired. To exhibit the effectiveness of SFLL-HD against the approximate attack, we utilize the empirical results from [14]. Figure 9.4 plots the execution time of the SAT, AppSAT, and Double-DIP attack on the 32-bit s38417 circuit as a function of h. It can be observed that the Double-DIP attack is applicable only when $h = 0$; for $h > 0$, it behaves just like the SAT attack. The AppSAT attack is effective for only small h values, i.e. $h \le 4$, and then behaves similarly to the SAT attack.

9.3.3   Resilience Trade-Offs

SFLL-HD allows a designer to use the Hamming distance h as a lever to trade resilience against one attack for resilience to another. Values of h closer to either 0 or k deliver higher resilience to the SAT attack, whereas resilience to the removal attack can be maximized by setting $h = k/2$. Figure 9.5 illustrates this trade-off between the removal attack resilience (in terms of the number of protected input patterns) and the security level s against the SAT attack for five benchmark circuits [14]. It can be observed that for SFLL-HD, the security-level s attained against SAT attacks varies polynomially with h ($h \in [0, k]$); the larger the number of protected patterns,

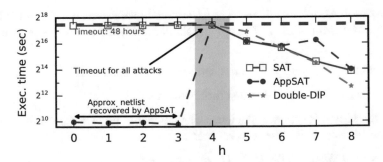

Fig. 9.4 Execution time of the SAT, AppSAT, and Double-DIP attack on (k=) 32-bit s38417 circuit plotted as a function of h. 1000 random queries are applied after every 12 AppSAT iterations. In the shaded region, all the three attacks time out. Double-DIP is applicable only for h = 0; for h > 0, it behaves the same as the SAT attack

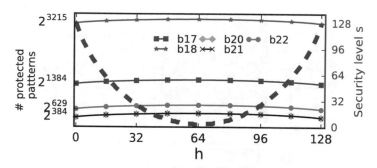

Fig. 9.5 SAT attack resilience vs. removal attack resilience for SFLL-HD; $k = 128$

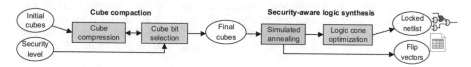

Fig. 9.6 The SFLL-flex algorithm has two main stages, cube compaction, and security-aware logic synthesis. The cube compaction stage compresses the user-specified input cubes and reduces the on-chip storage requirements. The security-aware logic synthesis cost-effectively strips functionality from the original circuit based on the compressed cubes

the lower the security level. The security level depends only on k and h, irrespective of the circuit. For the maximum number of protected patterns, i.e., $h = n/2$, the security level s is minimal. The security level is at its maximum at $h = 0$ or $h = k$.

In summary, among all logic locking techniques discussed in this book, SFLL-HD offers the most comprehensive protection against different classes of attacks.

9.4 SFLL-Flex

In contrast to SFLL-HD, SFLL-flex allows the designer to specify, and thus, protect the IP-critical input patterns. SFLL-flex is more suited for applications where the designer intends to protect a range of values that can be compactly represented using a set of input cubes. Examples include processors with to-be-protected address spaces, for which access may be granted to restricted entities [1]; network-on-chip (NoC) routers where specific IP address ranges may carry particular semantics [3]; intrusion detection systems that rely on pattern matching [4]; and digital signal processing applications, such as comb filters [2], which accentuate/attenuate frequencies at regular intervals that are multiples of a fundamental frequency. The restore unit in SFLL-flex stores the protected input patterns in a compact form, i.e., in the form of c input cubes, each with k specified bits. The input cubes can be conceived as the secret keys to be loaded onto the chip for the restore unit to recover the stripped functionality.

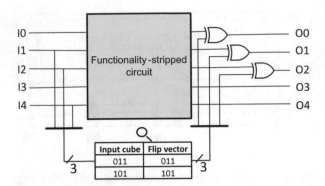

Fig. 9.7 SFLL-flex for a circuit with five inputs and five outputs. k = 3 in the example as only three bits of the protected input cubes are specified by the designer. f = 3 as three outputs are protected. The circuit is re-synthesized to strip the functionality from the original design based on the input cubes; these cubes are used along with the flip vectors to restore the original functionality

In a design with multiple outputs, not every output needs protection; only the IP-critical part of the design may be protected to control the cost of logic locking, which is at the discretion of the designer. SFLL-flex assists a designer in making such choices by providing him/her with an optimization framework that allows him/her to optimize the cost of functionality-stripping under the specified security constraints. SFLL-flex first compresses the protected input cubes into a compact representation and then enables the outputs of a circuit to be selectively flipped (and restored) for the protected input cubes; a *flip vector* associated with each protected input cube holds information regarding which outputs are to be flipped for the protected input cube. The framework presented in Fig. 9.6 has consists of two main stages: (1) a cube compaction stage, and (2) a security-aware logic synthesis stage. Further details are presented in Sect. 9.4.2.

9.4.1   Architecture

The restore unit of SFLL-flex consists of a tamper-proof [10] look-up table and XOR gates. The LUT stores c k-bit input cubes along with the corresponding f-bit flip vectors (for protecting f out of m outputs) that dictate the functionality stripped from the circuit. When the input matches an entry in the LUT, the associated flip vector is retrieved from the table and XORed with the outputs to restore the original functionality.

Example Figure 9.7 depicts the architecture of SFLL-flex with the locked circuit consisting of the FSC and the restore unit. The FSC differs from the original circuit for two protected input cubes $x01x1$ and $x10x1$, collectively representing eight input patterns. The restore unit stores the two input cubes and the corresponding flip vectors. In this example, only three out of five outputs are protected.

Implementation Cost The cost of SFLL-flex is proportional to the size of the LUT, in addition to f XOR gates inserted at the outputs of the FSC. The cost of the LUT can be denoted as $c \times (k + f)$, where f is a designer-defined parameter denoting the number of protected outputs. Cost minimization requires the minimization of c and k. Note that functionality-strip operation presents an opportunity for reducing implementation cost as will be discussed. The original circuit may be modified in a way that the FSC incurs lesser area or power compared to the original circuit. Thus, the overall cost of the SFLL-flex circuit is the cost of the LUT minus the savings obtained during the functionality-strip operation.

9.4.2 Optimization Framework

Given a desired security level s and a set of input cubes (or input patterns) C_{init} to be protected, both provided by the designer for a netlist N, SFLL-flex attempts to minimize the implementation cost: $Cost_{sf} + c \times k$, where $Cost_{sf}$ is the implementation cost of N_{sf}, the functionality-stripped netlist, and $c \times k$ is the implementation cost of the LUT. The optimization problem can be formulated as:

$$\text{minimize} \quad Cost_{sf} + c \times k \quad \text{such that} \quad k - \lceil \log_2 c \rceil \geq s$$

where $k - \lceil \log_2 c \rceil$ is the security level attained against the SAT attack. SFLL-flex solves this optimization problem by dividing it into two main stages [14]. In the first stage, it only focuses on compressing the input cubes (or input patterns) to minimize the LUT cost ($c \times k$). This stage generates the compactly represented final cubes C (or secret keys) that also satisfy the security constraint $k - \lceil \log_2 c \rceil \geq s$. The second stage re-synthesizes the logic of the protected outputs based on the keys obtained from the first stage with the goal of minimizing $Cost_{sf}$.

Stage 1: Cube Compression This objective of this stage to reduce the LUT cost, which is denoted as $c \times k$ and forms a major component of the overall implementation cost, i.e.,

$$\text{minimize} \quad c \times k \quad \text{such that} \quad k - \lceil \log_2 c \rceil \geq s$$

SFLL-flex assumes that the compacted cubes/keys lead to modifying at least one circuit output for at least one input pattern in every cube in C_{init}. Thus, the overall cube compression problem reduces to finding the minimal set of cubes that collectively intersect each cube in C_{init}. Algorithm 9 describes the heuristic approach utilized for solving this problem. The first step of the algorithm is recursive cube compression wherein compatible cubes are merged to minimize c, e.g., the cubes $0x100$ and $0110x$ can be merged into a single cube 01100.

Let us assume that a designer specifies the desired security level $s = 128$ and 15 initial cubes, each having 256 bits. Note that not all 256 bits are essential for the

---

Algorithm 9: Cube compression algorithm

Input : Initial cubes C_{init}, Security level s
Output: Final cubes C
1 $C \leftarrow$ merge_compatible_cubes(C_{init})
2 $s_{new} \leftarrow k - \log_2 c$
3 **while** $s_{new} \geq s$ **do**
4 | $C \leftarrow$ eliminate_conflicting_bit(C)
5 | $C \leftarrow$ merge_compatible_cubes(C)
6 | $s_{new} \leftarrow$ update_security_level(c, k)
7 **end**

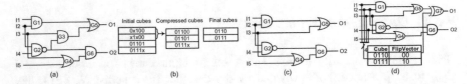

Fig. 9.8 Application of SFLL-flex to c17 ISCAS circuit [14]. (**a**) Original circuit. (**b**) Cube compression. (**c**) FSC. (**d**) Locked circuit

achieving desired security level $s = k - \lceil \log_2 c \rceil$. In this particular example, 132 bits are sufficient to satisfy the constraint $k - \lceil \log_2 15 \rceil \geq 128$. The second step of the cube compaction stage is to eliminate (equivalently, turn into x's) the bits that are conflicting among the cubes, while adhering to security level s. This step may further reduce c, as certain cubes become compatible to be merged.

Example Consider the c17 ISCAS benchmark circuit presented in Fig. 9.8, a set of four 5-bit initial cubes, and a security level $s = 3$, specified by the designer. The two initial cubes $0x100$ and $x1x00$ can be merged into one cube 01100, reducing c to three. Next, we can reduce k to four by eliminating the rightmost bit in all the cubes. Elimination of bits in conflict also leads to further reduction in c to two, as more cubes can now be merged; the achieved security level $s = 3$. Thus, compared to the initial $4 \times 5 = 20$ bits, only $2 \times 4 = 8$ bits need to be stored on-chip.

Stage 2: Security-Aware Logic Synthesis SFLL-flex assumes that a designer only specifies the input cubes C and the security level s, and leaves it to the logic locking algorithm to determine the flip vectors, with each vector specifying the outputs that need to be flipped/modified for a particular cube. SFLL- utilizes a security-aware synthesis algorithm that minimizes implementation cost of the functionality-stripped netlist N_{sf} without compromising security.

The security-aware synthesis, depicted in Algorithm 10, is a simulated annealing-based approach to determine the flip-vectors V that yields an area-optimal functionality-stripped netlist N_{sf}. In each iteration, a new solution (netlist) N_{new} is generated by changing a random bit in the flip vector V, which leads to inclusion/exclusion of the corresponding cube for a particular output. The solution

---

Algorithm 10: Security-aware synthesis algorithm

Input   : Original netlist N, Final cubes C
Output: Functionality-stripped netlist N_{sf}, Flip vector V
1 $V \leftarrow$ init_flip_vector(N)
2 $N_{sf} \leftarrow$ rand_soln(N, C)
3 $cost_{sf} \leftarrow$ cost(N_{sf})
4 $T = 1.0, T_{min} = 0.00001, \alpha = 0.9$
5 **while** $T > T_{min}$ **do**
6   **for** $i = 1$ *to* 200 **do**
7          $N_{new} \leftarrow$ neighbor(N_{sf}, C)
8          $cost_{new} \leftarrow$ cost(N_{new})
9   **if** $Rand(0, 1) < exp(\frac{cost_{new} - cost_{sf}}{T})$ **then**
10             $N_{sf} \leftarrow N_{new}$
11             $cost_{sf} \leftarrow cost_{new}$
12             $V \leftarrow$ update_flip_vector(N_{sf}, po)
13   **end**
14   **end**
15     $T = T \times \alpha$
16 **end**

N_{new} is retained if it yields cost savings, i.e. $cost_{new} < cost_{sf}$. Occasionally, an inferior solution may be accepted, with a probability of $exp(\frac{cost_{opt} - cost_{new}}{T})$, to prevent the simulated annealing algorithm from getting stuck at a local optimum.

Example Figure 9.8 illustrates the application of security-aware synthesis to the c17 circuit. Algorithm 10 takes as input (1) the original c17 netlist and (2) the final cubes (produced by Algorithm 9), and generates the FSC. As part of the optimization process, the gate G3 is removed from the logic cone O1, flipping its output for the cube 0110x and potentially leading to area/power savings. The associated flip vector 10 restores the correct output for logic cone O1 by flipping "back" its output for the cube 0110x.

9.4.3   Security Analysis

SAT Attack Resilience An attacker, following either a SAT-based or a random guess-based attack model, must identify *all* input patterns of the protected input cubes in SFLL-flex to be able to recover the correct functionality of the original design from the on-chip implementation. In contrast to SFLL-HD, the protected input cubes can be arbitrary in SFLL-flex, and one cube does not infer another. This requires the retrieval of the content of the entire LUT that represents the stripped functionality. Nevertheless, in this section, the security of SFLL is measured conservatively; the attack success is defined by the attacker's ability to retrieve any input pattern that belongs to one of the protected input cubes.

Theorem 9.7 *SFLL-flex is $(k - \lceil \log_2 c \rceil)$-secure against SAT attack.*

Proof For SFLL-flex, the cardinality of the set of protected cubes P is $|P| = c$. Thus, from Eq. 9.1, the success probability of a PPT adversary making a polynomial number of queries $q(k)$ is given by

$$
\begin{aligned}
P_{succ} &= \frac{|P|}{2^k} + \frac{|P|}{2^k - 1} \cdots \frac{|P|}{2^k - q(k)} \\
&= \frac{c}{2^k} + \frac{c}{2^k - 1} \cdots \frac{c}{2^k - q(k)} \\
&\approx \frac{q(k) \cdot c}{2^k} \\
&< \frac{q(k)}{2^{k - \lceil \log_2 c \rceil}}
\end{aligned}
$$

Thus, SFLL-flex is $(k - \lceil \log_2 c \rceil)$-secure against the SAT attack.

Sensitization Attack Resilience The analysis is similar to that for SFLL-HD.

Theorem 9.8 *SFLL-flex is k-secure against sensitization attack.*

Proof All the k bits of SFLL-flex converge within the comparator inside the LUT to produce the signal that asserts the XOR vector operation between the flip vector and the outputs. Therefore, sensitizing any key bit through the LUT to any of the outputs requires controlling all the other key bits. All k bits are therefore pairwise-secure. SFLL-flex is k-secure against sensitization attack.

Removal Attack Resilience Removal attack on SFLL-flex involves isolating the restore circuit, comprising and LUT and a few XOR gates, from the rest of the circuit.

Theorem 9.9 *SFLL-flex is $c \cdot 2^{n-k}$-resilient against removal attack.*

Proof Even if the LUT along with its surrounding logic can be identified by a reverse-engineer, he/she can only recover the FSC denoted as N_{rec}. However, N_{rec} produces incorrect output for the protected input patterns Γ. Thus, similar to Theorem 9.3,

$$
N_{rec}(i) \neq F(i), \quad \forall i \in \Gamma
$$

$$
\begin{aligned}
|\Gamma| &= |P| \times 2^{n-k} \\
&= c \cdot 2^{n-k}
\end{aligned}
$$

So, SFLL-flex is $c \cdot 2^{n-k}$-resilient against a removal attack.

Table 9.1 Comparative security analysis of logic locking techniques against existing attacks

Attack/defense	Anti-SAT [11]	SARLock [12]	TTLock [15]	SFLL-HD [14]	SFLL-flex [14]
SAT	k	k	k	$k - \lceil \log_2 \binom{k}{h} \rceil$	$(k - \lceil \log_2 c \rceil)$
Sensitization	k	k	k	k	k
Removal	0	0	2^{n-k}	$\binom{k}{h} \cdot 2^{n-k}$	$c \cdot 2^{n-k}$
Approx	$p/2^k$	$1/2^k$	$2/2^k$	$\binom{k}{h}/2^k$	$c/2^k$

SFLL is secure against all attacks. Various versions of SFLL offer a trade-off between SAT attack resilience and removal attack resilience

As these theorems show, the number and the size of the protected input cubes, denoted by c and k respectively, dictate the trade-off between resilience to oracle-guided and removal attacks.

Approximate Attack Resilience The OER obtained by SFLL-flex is dictated by c. Since, at least one output will be flipped for each c, we can expect that for an arbitrary key value, there will be c incorrect entries in the corresponding column, leading to an effective OER of $\frac{c}{2^k}$.

In summary, SFLL outperforms all existing logic locking techniques in terms of achieving holistic security against different classes of attacks. In addition to the better security it offers, SFLL allows a designer the flexibility to specify which parts of a circuit he/she intends to protect. SFLL also enables trade-offs between SAT attack resilience and removal attack resilience. While Table 2.1 presents an attack/defense matrix for all logic locking techniques, Table 9.1 presents a more detailed comparison of SFLL with the other SAT attack resilient techniques. SFLL, however, relies on existing logic synthesis tools to effect the functionality-strip operation. These EDA tools are not designed to take security into account and might leave traces in a netlist that help attackers retrieve the original functionality. Recently, Sengupta et al. [6] have developed SFLL_fault, which is a secure method for effecting the functionality-strip operation in SFLL-flex by making use of automatic test pattern generation. SFLL-HD currently does not offer any security guarantees regarding the functionality-strip operation. Developing provably secure methods for generating a functionality-stripped netlist is still an open research problem.

References

1. Chen Q, Azab AM, Ganesh G, Ning P (2017) Privwatcher: non-bypassable monitoring and protection of process credentials from memory corruption attacks. In: ACM Asia conference on computer and communications security, pp 167–178
2. Chu S, Burrus C (1984) Multirate filter designs using comb filters. IEEE Trans Circuits Syst 31(11):913–924
3. Diguet J, Evain S, Vaslin R, Gogniat G, Juin E (2007) NOC-centric security of reconfigurable soc. In: IEEE first international symposium on networks-on-chip, pp 223–232

4. Kumar S, Dharmapurikar S, Yu F, Crowley P, Turner J (2006) Algorithms to accelerate multiple regular expressions matching for deep packet inspection. In: ACM SIGCOMM computer communication review, pp 339–350
5. Rajendran J, Pino Y, Sinanoglu O, Karri R (2012) Security analysis of logic obfuscation. In: IEEE/ACM design automation conference, pp 83–89
6. Sengupta A, Nabeel M, Yasin M, Sinanoglu O (1996) ATPG-based cost-effective, secure logic locking. In: IEEE VLSI test symposium. IEEE, Piscataway, pp 2–8
7. Shamsi K, Li M, Meade T, Zhao Z, Pan DZ, Jin Y (2017) AppSAT: approximately deobfuscating integrated circuits. In: IEEE international symposium on hardware oriented security and trust, pp 95–100
8. Shen Y, Zhou H (2017) Double DIP: re-evaluating security of logic encryption algorithms. Cryptology ePrint Archive, Report 2017/290. http://eprint.iacr.org/2017/290
9. Subramanyan P, Ray S, Malik S (2015) Evaluating the security of logic encryption algorithms. In: IEEE international symposium on hardware oriented security and trust, pp 137–143
10. Tuyls P, Schrijen G, Škorić B, van Geloven J, Verhaegh N, Wolters R (2006) Read-proof hardware from protective coatings. In: Goubin L, Matsui M (eds) International conference on cryptographic hardware and embedded systems, pp 369–383
11. Xie Y, Srivastava A (2016) Mitigating SAT attack on logic locking. In: International conference on cryptographic hardware and embedded systems, pp 127–146
12. Yasin M, Mazumdar B, Rajendran J, Sinanoglu O (2016) SARLock: SAT attack resistant logic locking. In: IEEE international symposium on hardware oriented security and trust, pp 236–241
13. Yasin M, Mazumdar B, Sinanoglu O, Rajendran J (2017) Removal attacks on logic locking and camouflaging techniques. IEEE Trans Emerg Top Comput 99(0):PP
14. Yasin M, Sengupta A, Nabeel MT, Ashraf M, Rajendran J, Sinanoglu O (2017) Provably-secure logic locking: from theory to practice. In: ACM/SIGSAC conference on computer & communications security, pp 1601–1618
15. Yasin M, Sengupta A, Schafer B, Makris Y, Sinanoglu O, Rajendran J (2017) What to lock?: functional and parametric locking. In: Great lakes symposium on VLSI, pp 351–356

Chapter 10
Side-Channel Attacks

Abstract Apart from the previously mentioned attacks that exploit the algorithmic weaknesses of logic locking techniques, logic locking is also vulnerable to the emerging class of side-channel attacks, which are the focus of this chapter. The chapter introduces four attacks on logic locking that exploit various side-channels to extract secret key. The differential power analysis attack utilizes the power consumption of a chip to determine the secret logic locking key. The test-data mining attack and the hill climbing attack determine the secret key from the test data. The de-synthesis attack extracts the key by leveraging the traces left in a netlist during logic synthesis.

The attacks discussed so far in this book aim at exploiting the algorithmic weaknesses of logic locking techniques. However, secret information may also be leaked through side-channels such as power, electromagnetic radiation, and time [3, 4]. This chapter presents four representative side-channel attacks on logic locking. Section 10.1 presents the differential power analysis (DPA) attack on logic locking. Apart from the traditional side-channels such as timing and power, logic locking has also been shown to be vulnerable to newer classes of side-channel attacks that leverage the vulnerabilities associated with different stages of the IC design flow. Sections 10.2 and 10.3 present the test-data mining (TDM) attack and the hill climbing attack, respectively; both attacks exploit test data to extract sensitive information. Section 10.4 introduces the de-synthesis attack that derives the secret key from the information embedded in a netlist during logic synthesis.

10.1 Differential Power Analysis (DPA) Attack

10.1.1 Basic Idea

Power side-channel attacks exploit the correlation between the power consumption of the CMOS logic and the data being processed [3]. Many such attacks have successfully utilized the power measurements to extract the secret keys for multiple

© Springer Nature Switzerland AG 2020   119
M. Yasin et al., *Trustworthy Hardware Design: Combinational Logic
Locking Techniques*, Analog Circuits and Signal Processing,
https://doi.org/10.1007/978-3-030-15334-2_10

cryptographic algorithms. The DPA attack is the most powerful variant of the power side-channel attacks [3]. It has been demonstrated that DPA attack is also applicable to logic locking [8]. The attack can be mounted on the basic logic locking techniques in a divide-and-conquer fashion, i.e., by targeting the logic cones on an individual basis. Section 10.1.2 introduces the basic concepts of a DPA attack, and Sect. 10.1.3 elaborates on the application of DPA to logic locking.

10.1.2 Preliminaries: The DPA Attack

A DPA attack that retrieves the secret key of a cryptographic algorithm comprises the following two major steps [3]: (1) power trace acquisition, and (2) statistical analysis.

Power Trace Acquisition The DPA attack starts by carrying out m encryption operations by supplying m random plaintexts $P_{1 \cdots m}$ to the target IC and recording the corresponding ciphertexts $C_{1 \cdots m}$ as well as the associated power traces $T_{1 \cdots m}[1 \cdots k]$, with each power trace containing k samples.

Statistical Analysis This offline computation step classifies the acquired power traces into two classes based on a selection function $D(C, b, K_s)$, where b represents an intermediate bit b (e.g., an output of a round), K_s denotes a key guess by the attacker, and C denotes the ciphertext. Depending on the value of b, the power traces are classified into a zero-bin T_0 (if $b = 0$) and a one-bin T_1 (if $b = 1$). The absolute difference of the average of the power traces stored in the two bins $\Delta_D[j]$ is then computed as follows:

$$\Delta_D[j] = \left| \frac{1}{|T_0|} \sum_{T_i[j] \in T_0} T_i[j] - \frac{1}{|T_1|} \sum_{T_i[j] \in T_1} T_i[j] \right| \tag{10.1}$$

$$T_0 = \{T_i[j] | D(C, b, K_s)) = 0\}$$
$$T_1 = \{T_i[j] | D(C, b, K_s)) = 1\} \tag{10.2}$$
$$1 \leq j \leq k$$

For an incorrect key guess, the computed value of the bit b will differ from its correct value, in about half of the instances, making the selection function uncorrelated with the actual data. Roughly half of the power traces go to each bin, rendering $\Delta_D[j]$ close to zero. For the correct key, however, the distribution of 1's and 0's computed by the selection function will perfectly match the distribution exhibited by the actual chip data. The bins will no longer be populated randomly, resulting in a spike in the differential trace $\Delta_D[j]$. This characteristic of the differential trace has been studied extensively to break many cryptographic algorithms.

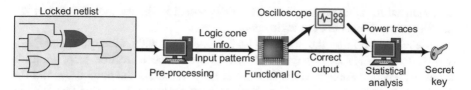

Fig. 10.1 Differential power analysis attack on logic locking

Threat Model The threat model for the DPA attack is similar to that of SAT attack; the attacker has access to a locked netlist and a functional IC, which can also be instrumented for collecting the power traces.

10.1.3 DPA Attack on Logic Locking

The goal of the DPA attack on a logic locked chip is to reveal the secret key used to lock the IC. An overview of the attack is shown in Fig. 10.1. The attack targets individual logic cones in a divide-and-conquer approach. For each logic cone, the attacker generates the primary input patterns and applies them to the functional IC. An oscilloscope monitors one of the I/O pins and captures the circuit output along with the power traces as the patterns are applied. This is followed by statistical analysis on the collected traces that determines the correct key for the targeted logic cone.

As depicted in Algorithm 11, the first step of the DPA attack on logic locking is to parse the locked netlist N_{lock} to determine the key inputs KI, the primary inputs PI, and the primary outputs PO. The set of recovered key bits $K_{resolved}$

Algorithm 11: DPA attack on a logic locked netlist

Input: Locked netlist N_{lock}, Functional IC F
Output: Set of resolved key values $K_{resolved}$
1 $KI, PI, PO \leftarrow$ parse_locked_netlist(N_{lock})
2 $K_{resolved} \leftarrow \emptyset$
3 $PO \leftarrow$ sort_with_KI(PO)
4 **for** *each* PO_{cur} *in* PO **do**
5 $\quad$ $KI_{cur}, PI_{cur}, C_{cur} \leftarrow$ identify_logic_cone(N_{lock}, PO_{cur})
6 $\quad$ $O_F, PT_F \leftarrow$ apply_patterns(C_{cur}, PI_{cur})
7 $\quad$ $K_{DPA} \leftarrow$ statistical_analysis($C_{cur}, PT_F, KI_{cur}, PI_{cur}, K_{resolved}$)
8 $\quad$ **if** *verify($C_{cur}, K_{DPA}, I_{cur}, O_F$)* **then**
9 $\quad\quad$ $K_{resolved} \leftarrow K_{resolved} \cup K_{DPA}$
10 $\quad\quad$ **if** $|K_{resolved}| = k$ **then**
11 $\quad\quad\quad$ break
12 $\quad\quad$ **end**
13 $\quad$ **end**
14 **end**
15 **return** $K_{resolved}$

---

is initialized to an empty set. The set PO is sorted in the ascending order of the number of key inputs in each logic cone.

In each iteration of the attack, a logic cone C_{cur} with primary inputs PI_{cur} and key inputs KI_{cur} is targeted. In the power trace collection step, multiple input patterns are applied to the functional IC F to obtain the output O_F and the associated power traces PT_F. The objective of the statistical analysis that follows is to recover the correct key values K_{DPA} for the logic cone under attack. This step involves simulating the logic cone C_{cur} with multiple guesses for the key value and computing the selection function O_{cur}. Based on the value O_{cur}, the power traces PT_F are added to either bin T_0 or T_1, as follows:

$$T_0 = \{PT_F | O_{cur}(KI_{cur}; PI_{cur}) = 0\}$$
$$T_1 = \{PT_F | O_{cur}(KI_{cur}; PI_{cur}) = 1\}$$

(10.3)

Equation 10.3 is computed over the entire set of primary inputs PI_{cur}. The differential trace Δ_{cur} for the two bins is computed as:

$$\Delta_{cur} = \left| \frac{1}{|T_0|} \sum_{PT_F \in T_0} PT_F - \frac{1}{|T_1|} \sum_{PT_F \in T_1} PT_F \right|$$

(10.4)

For each guess of KI_{cur}, a value for Δ_{cur} is obtained. The key guess that yields the maximum value for Δ_{cur} is determined as K_{DPA}. To verify the correctness of the K_{DPA}, the logic cone C_{cur} is driven with K_{DPA} and all possible value of PI_{cur}; if the observed output matches with O_F, all the key bits in K_{DPA} are considered as resolved and are added to the set $K_{resolved}$. The attack on individual logic cones is carried out until all the key bits have been resolved (i.e., $|K_{resolved}| = k$) or all the logic cones have been processed.

The DPA attack results reported in [8] demonstrate that the attack is highly successful against RLL. However, the attack loses its effectiveness against SLL. The resilience of SLL to DPA can be attributed to (1) a large number of key/primary inputs in individual logic cones, rendering the attack computationally harder, and (2) a phenomenon referred to as key aliasing, where multiple key values exhibit the same Δ_{cur} (refer to [8] for details). The DPA attack is also ineffective against point function-based logic locking, since (1) the circuit cannot be divided into smaller logic cones (with only a handful of key inputs), and (2) the Δ_{cur} is almost identical for the correct as well as incorrect key values due to the extremely low output corruption exhibited by the point function-based locking techniques.

10.2 Test-Data Mining (TDM) Attack

10.2.1 Basic Idea

Each fabricated chip goes through a manufacturing test that screens out the defective chips [1]. The test may be conducted in the foundry or in an outsourced assembly

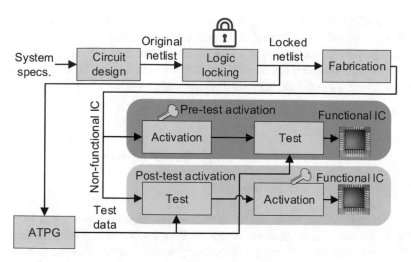

Fig. 10.2 Pre-test and post-test activation models for logic locking

and test facility (OSAT). The test is conducted using automated test equipment (ATE), which applies test patterns to an IC and compares the IC output with the expected response. The test patterns and the expected responses are computed using ATPG algorithms and are sent to the test facility.

In a logic locked IC, each key input is connected to an on-chip tamper-proof memory bit. An IC is activated by loading the secret key to the on-chip memory. An ATPG algorithm must take into account the values that will be applied to the key inputs during the manufacturing test. As illustrated in Fig. 10.2, existing logic locking schemes follow two activation models: pre-test and post-test activation [6, 9]. In pre-test activation, the secret is loaded onto the chip prior to the manufacturing test. In post-test activation, the manufacturing test is conducted using dummy key values, and the chip is activated post-test.

The TDM attack targets the pre-test activation model. In this model, both manufacturing test and ATPG are conducted with the secret key in place. During ATPG, the secret key is used as a constraint on the key inputs; thus, the generated test stimuli and responses may reveal information about the secret key.

Threat Model The TDM attack does not require a functional IC. The attacker requires only (1) the locked netlist N_{lock}, and (2) the test stimuli T and responses Γ.

10.2.2   TDM Attack Algorithm

During test pattern generation for a locked netlist N_{lock}, the secret key K_s is applied as a constraint and a set of test patterns that maximizes fault coverage is obtained. Inevitably, information about the secret key may be revealed by the test patterns, since the ATPG tool has to satisfy the key value constraints. The TDM attack simply solves the "reverse ATPG" problem; it uses the test patterns as constraints and tries to find a potentially correct key K_P that maximizes the fault coverage. The attack is an optimization problem: the objective is to maximize the fault coverage FC under the test stimulus T and test response Γ constraints, as follows:

$$
\begin{aligned}
\text{maximize} \quad & FC \\
\text{subject to} \quad & \underset{1 \leq i \leq N}{\forall} N_{lock}(K_P, T_i) = \Gamma_i \qquad (10.5) \\
\text{solve for} \quad & K_P
\end{aligned}
$$

Equation 10.5 formulates a system of Boolean equations which can be solved using techniques such as Boolean satisfiability or Integer Linear Programming [2]. ATPG algorithms are also capable of solving such a system of Boolean equations while at the same time maximizing the fault coverage. In Eq. 10.5, the fault coverage for the locked netlist is maximized subject to N test stimuli and response constraints. As illustrated in Fig. 10.4, one of the options to achieve this objective is to instantiate N replicas of N_{lock}. All the instances share the common key inputs; however, the inputs and outputs of the ith instance N_{lock_i} are subjected to T_i and Γ_i constraints, respectively. The ATPG tool applies the specified constraints and extracts the key value that maximizes the fault coverage.

Example Let us consider the locked netlist shown in Fig. 10.3. When the correct key value $K_s = 00$ is used as a constraint during test pattern generation, eight test patterns are generated by the ATPG tool as listed in Table 10.1. When the TDM attack (illustrated in Fig. 10.4) is launched on the netlist, the only key that maximizes the fault coverage and satisfies the test pattern constraints is 00; the corresponding fault coverage is 82.43%. In this particular example, none of the other key values satisfies the test pattern constraints.

Fig. 10.3 A netlist locked with two key gates [7]. The correct key is 00

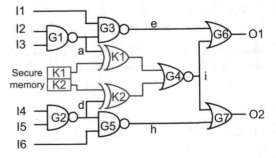

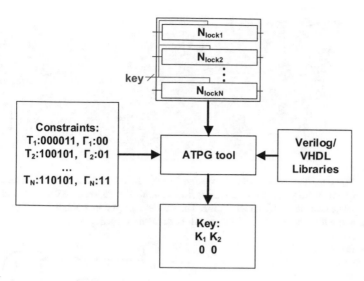

Fig. 10.4 Test-data mining attack on pre-test activation model

10.2.3 HackTest Attack on IC Camouflaging

The TDM attack is also applicable in the context of IC camouflaging [10]. HackTest, the camouflaging counterpart of TDM attack, can be launched to extract the correct functionality of the camouflaged gates, as depicted in Fig. 10.5. The HackTest attack first models each camouflaged gate using a multiplexer whose select lines serve as key inputs, and then launches the TDM attack on the logic locked equivalent netlist. In the camouflaged netlist shown in Fig. 10.6a—each camouflaged gate implements either a NAND or a NOR. Each camouflaged gate is modeled using a MUX as in Fig. 10.6b. The TDM and HackTest attacks are also applicable in the presence of scan compression (refer to [10] for details).

Table 10.1 Pre-test activation test patterns for the netlist in Fig. 10.3

Key(K_s)	Stimulus (T)	Response (Γ)
00	011001	10
00	101010	01
00	101111	01
00	011101	10
00	111010	11
00	000111	11
00	110001	00
00	001011	10

The correct key $K_s = 00$ is used as a constraint during ATPG

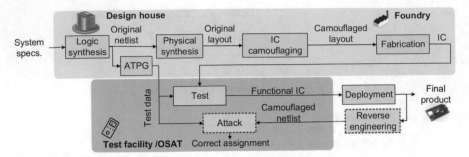

Fig. 10.5 The HackTest attack flow. The designer sends the test data to the test facility for use in manufacturing test. An adversary in the test facility can misuse the legitimate access to test data to compromise the security of IC camouflaging. The trusted design house and test facility are shaded green; the untrusted test facility/OSAT and end-user are shaded red

Fig. 10.6 (**a**) A camouflaged netlist with each camouflaged gate implementing either a NAND or a NOR; the true functionality of the gate is highlighted in red. (**b**) Modeling the logic locked equivalent of the camouflaged netlist using multiplexers. Each camouflaged gate is replaced with a multiplexer; the value of the select lines decides the functionality the gate takes on [10]

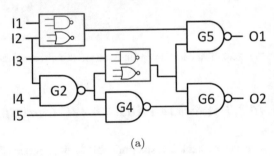

(a)

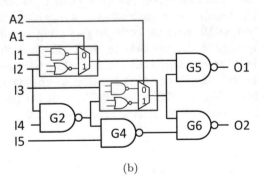

(b)

The TDM and HackTest attacks raise serious concerns about security in the emerging business model of OSATs. Today, foundries may outsource test and assembly services to OSATs. With access to critical assets such as test data, malicious agents in the OSATs may leak sensitive information.

10.3 Hill Climbing Search Attack

10.3.1 Basic Idea

Similar to the TDM attack, the hill climbing attack makes use of test data to retrieve the secret key under the same threat model [6]. The attack relies on the assumption that upon applying test stimuli to a locked netlist, the netlist output will match the expected test responses for the correct key. However, for an incorrect key, the netlist output will exhibit significant dissimilarity from the expected test responses. The dissimilarity can be represented in terms of the Hamming distance between the test responses and the netlist output.

Threat Model Similar to the TDM attack, the hill climbing does not require a functional IC. The attacker requires only (1) the locked netlist N_{lock}, and (2) the test stimuli T and responses Γ.

10.3.2 Attack Algorithm

As depicted in Algorithm 12, the hill climbing attack relies on a hill climbing-based approach to recover the secret key K_{rec} [6]. The attack starts by making a random initial guess for K_{rec}. The output R_A of the locked netlist N_{lock} is recorded for the test stimuli T and the key K_{rec}. A randomly selected key bit in K_{rec} is then toggled and the output R_B is recorded. The Hamming distance of both R_A and R_B to the test response Γ are computed. The toggle is retained if R_B exhibits a smaller Hamming distance compared to that for R_A. The attack is successful when a key value K_{rec} is found that leads to a Hamming distance of zero. The correctness of the recovered key can be more exhaustively verified if a functional IC is available. The toggle operation may be repeated for all the key bits in K_{rec}. Moreover, several initial guesses for the secret key may be made until one leads to a successful attack or produces a Hamming distance value very close to zero [6].

While the hill climbing attack is effective against RLL and FLL, it loses its effectiveness against SLL [9]. The TDM attack, however, is equally effective against all three basic logic locking techniques. Furthermore, the effectiveness of the hill climbing attack in the presence of scan compression still needs to be established. As a countermeasure against the attack, test-aware logic locking technique may be utilized, which inserts MUX key gates into a design that decouple the test responses from the key bit values [6].

Algorithm 12: Hill climbing attack algorithm [6]

Input : Locked netlist N_{lock},Test stimuli T, Test responses Γ
Output: Recovered key K_{rec}

1 found = False
2 **while** *!found* **do**
3      $K_{rec} \leftarrow$ random_key_guess()
4   **foreach** *random rkbit* $\in K_{rec}$ **do**
5          $R_A \leftarrow$ simulate_netlist (N_{lock}, K_{rec}, T)
6          $HD_A \leftarrow$ compute_hamming_distance (R_A, Γ)
7          $rkbit \leftarrow !rkbit$
8          $R_B \leftarrow$ simulate_netlist (N_{lock}, K_{rec}, T)
9          $HD \leftarrow$ compute_hamming_distance (R_B, Γ)
10   **if** $HD < HD_A$ **then**
11              $rkbit \leftarrow !rkbit$
12              $HD \leftarrow HD_A$
13   **end**
14   **if** $HD == 0$ **then**
15   **if** *verify* (N_{lock}, K_{rec}) **then**
16   found $\leftarrow$ True
17   **end**
18   break
19   **end**
20   **end**
21 **end**

10.4 De-synthesis Attack

10.4.1 Basic Idea

The basic locking techniques typically lock a gate level netlist, which is generated by logic synthesis. These techniques insert XOR/XNOR and other types of key gates into a netlist and may re-synthesize the locked netlist. Apart from the inserted key gates, the original and the locked netlist may be expected to be similar, in terms of the number and types of gates in a netlist. The de-synthesis attack relies on the observation that if a locked netlist is re-synthesized with an incorrect key, the resulting netlist may be significantly different from the locked netlist [5]. Only upon synthesis with the correct key or with a key only slightly different from the correct key, one can expect the locked and the re-synthesized netlists to be similar [5].

Threat Model The de-synthesis attack requires only the locked netlist N_{lock} to operate. It does not require a functional IC.

Algorithm 13: The de-synthesis attack algorithm [5]

    Input   : Locked netlist N_{lock}, Number of random guesses Z
    Output: Recovered key K_{rec}
1 $abs\_min = \infty$
2 **for** $i=1$ *to* Z **do**
3 | $K_r \leftarrow$ random_key_guess()
4 | **while** *true* **do**
5 | | $K_1 \leftarrow \underset{k:ham(K,K_r)\leq 1}{argmin} \Delta(N_{lock}, S(N_{lock}, K))$
6 | | **if** $K_1 == K_r$ **then**
7 | | | break
8 | | **end**
9 | | **else**
10 | | | $K_r = K_1$
11 | | **end**
12 | **end**
13 | $cur\_min \leftarrow \Delta(N_{lock}, S(N_{lock}, k_1))$
14 | **if** $cur\_min < abs\_min$ **then**
15 | | $K_{rec} \leftarrow K_1$
16 | | $abs\_min \leftarrow cur\_min$
17 | **end**
18 **end**

10.4.2   Attack Algorithm

Notation The objective of the de-synthesis attack is to retrieve a key k_{rec} such that the locked netlist N_{lock} is the most similar to $S(N_{lock}, k_{rec})$, i.e., the netlist obtained upon re-synthesizing N_{lock} using a synthesis algorithm S with the key value k_{rec} applied as a constraint on the key inputs. The dissimilarity $\Delta(C_1, C_2)$ of two netlists C_1 and C_2 can be measured as follows:

$$\Delta(C_1, C_2) = \sum_{i=1}^{T} (n_{g_i}[C_1] - n_{g_i}[C_2])^2 \tag{10.6}$$

where T denotes the cardinality of the set of gate types $\{g_1, g_2, \cdots, g_T\}$, and $n_G[C]$ denotes the count of gates of type G in the netlist C.

Attack Algorithm The approach that the de-synthesis attack algorithm takes, depicted in Algorithm 13, is very similar to the simulated annealing-based approaches used earlier in the context of SFLL-flex (Sect. 9.4) and the hill climbing attack (Sect. 10.3). The attack starts by making a random guess for the correct key k_r. The attack then re-synthesizes the locked netlist N_{lock} for all key values that are neighbours of K_r, i.e., differ from K_r in only one bit, as dictated by a Hamming distance *ham* of one. The key value K_1 that leads to the minimum dissimilarity between L and $S(N_{lock}, K_r)$ is chosen as the best guess for K_{rec}. The attack is repeated Z times, each time with a new random guess k_r, and the overall best guess for K_{rec} is determined.

The empirical results indicate that the de-synthesis attack can retrieve about 70% of key bits for RLL [5]. The effectiveness of the attack on point function-based locking and other SAT attack resilient techniques has not been established. The attack, however, demonstrates new vulnerabilities associated with logic locking.

In summary, this chapter presented four side-channel attacks on logic locking that make use of side-channels such as power consumption and test data to retrieve information about the secret key. The emergence of these attacks mandates that logic locking algorithms take into account the information leakage through side-channel attacks.

References

1. Bushnell M, Agrawal VD (2000) Essentials of electronic testing for digital, memory and mixed-signal VLSI circuits, vol 17. Springer Science and Business Media, New York
2. Clarke E, Gupta A, Kukula J, Strichman O (2002) SAT based abstraction-refinement using ILP and machine learning techniques. In: Computer aided verification. Springer, Berlin, pp 265–279
3. Kocher P, Jaffe J, Jun B (1999) Differential power analysis. In: Advances in cryptology. Springer, Berlin, pp 388–397
4. Kovatsch M, Duquennoy S, Dunkels A (2011) A low-power CoAP for Contiki. In: IEEE eighth international conference on mobile ad-hoc and sensor systems, pp 855–860
5. Massad M, Zhang J, Garg S, Tripunitara M (2017) Logic locking for secure outsourced chip fabrication: a new attack and provably secure defense mechanism. CoRR abs/1703.10187. http://arxiv.org/abs/1703.10187
6. Plaza S, Markov I (2015) Solving the third-shift problem in IC piracy with test-aware logic locking. IEEE Trans CAD Integr Circuits Syst 34(6):961–971
7. Rajendran J, Pino Y, Sinanoglu O, Karri R (2012) Security analysis of logic obfuscation. In: IEEE/ACM design automation conference, pp 83–89
8. Yasin M, Mazumdar B, Ali SS, Sinanoglu O (2015) Security analysis of logic encryption against the most effective side-channel attack: DPA. In: IEEE international symposium on defect and fault tolerance in VLSI and nanotechnology systems, pp 97–102
9. Yasin M, Saeed SM, Rajendran J, Sinanoglu O (2016) Activation of logic encrypted chips: pre-test or post-test? In: Design, automation test in Europe, pp 139–144
10. Yasin M, Sinanoglu O, Rajendran J (2017) Testing the trustworthiness of IC testing: an oracle-less attack on IC camouflaging. IEEE Trans Inf Forensics Secur 12(11):2668–2682

Chapter 11
Discussion

Abstract The last chapter of this book presents a summary of the logic locking defenses and attacks discussed throughout the book. The chapter also highlights the challenges faced by existing logic locking approaches and hints at the future research directions.

This chapter concludes the book by summarizing the techniques discussed throughout the book and offering directions for future research. Section 11.1 revisits the attack/defense matrix introduced in Chap. 2, summarizing the relation among various classes of attacks and defenses. Section 11.2 offers insights into the challenges faced by logic locking techniques. Section 11.3 highlights the future research directions.

11.1   Revisiting the Attack/Defense Matrix

Table 11.1 revisits the resilience of the existing logic locking techniques against different classes of attacks. The logic locking defenses are broadly classified as pre-SAT and post-SAT. The pre-SAT defenses represent various key gate selection algorithms [10–12]. Apart from being an easy target for the SAT attack, the pre-SAT defenses remain vulnerable to the approximate attacks and side-channel attacks. Among the post-SAT defenses, point function-based logic locking techniques remain the most common countermeasure against the SAT attack. These techniques, however, suffer from low output corruptibility. Among the other post-SAT defenses, cyclic logic locking [13] can be easily broken with CycSAT [31], whereas ORF incurs impractically high implementation overhead; moreover, secure synthesis of the one-way function and the original netlist remains an open research problem.

Among the attacks, the SAT attack stands out as the most powerful attack launched against logic locking. All the existing defenses against the attack exhibit various shortcomings that limit their use in practical settings. Certain techniques compromise on the output corruptibility, others incur high area, power, and timing

© Springer Nature Switzerland AG 2020
M. Yasin et al., *Trustworthy Hardware Design: Combinational Logic Locking Techniques*, Analog Circuits and Signal Processing,
https://doi.org/10.1007/978-3-030-15334-2_11

Table 11.1 Attack resiliency of logic locking techniques

| | Attack/defense | Pre-SAT | | | Post-SAT | | | Compound | Cyclic | ORF-Lock |
		RLL [12]	FLL [2, 11]	SLL [10]	AntiSAT [21]	SARLock [24]	SFLL [29]	[21, 24]	[13]	[26]
Algo.	Sensitization [10]	✗	✗	✓	✓	✓	✓	✓	✓	✓
	SAT [17]	✗	✗	✗	✓	✓	✓	✓	✓	✓
	CycSAT [31]	✗	✗	✗	✓	✓	✓	✓	✗	✓
App.	AppSAT [14]	✗	✗	✗	✓	✓	✓	✗*	✓	✓
	Double-DIP [15]	✗	✗	✗	✓	✓	✓	✗*	✓	✓
Removal	Removal/SPS [25]	✓	✓	✓	✗	✗	✓	✓	✓	✗
	AGR [28]	✓	✓	✓	✗	✗	✓	✗	✓	✓
	Bypass [22]	✓	✓	✓	✗	✗	✓	✓	✓	✓
Side chan.	Test-data mining [27]	✗	✗	✗	✓	✓	✓	✓	✓	✓
	Hill climbing [9]	✗	✗	✓	✓	✓	✓	✓	✓	✓
	DPA [23]	✗	✗	✗	✓	✓	✓	✓	✓	✓
	Desynthesis [6]	✗	✗	✗	✓	✓	✓	✓	✓	✓

✗ denotes that a technique is vulnerable the attack, ✓ denotes resilience against the attack, and ✗* denotes partial attack success

overhead, and a few require secure logic synthesis algorithms. Even SFLL-HD, the strongest logic locking defense to-date, trades off SAT attack resilience for output corruptibility and relies on the logic synthesis algorithms to generate a trace-free functionality stripped circuit. The existing approximate attacks mainly target compound logic locking techniques, whereas, the removal attacks focus mainly on detecting and removing the point functions. As for the side-channel attacks, they have mostly been explored in the context of pre-SAT logic locking techniques.

11.2 Challenges Faced by Logic Locking

In the light of the aforementioned discussion, we can enlist the following major challenges faced by logic locking techniques:

1. **Low output corruptibility**. Most SAT attack defenses compromise on the output corruptibility to achieve high SAT attack resilience. Consequently, the locked circuit rarely exhibits incorrect output, rendering logic locking infeasible for applications that demand high output corruptibility. Moreover, existing techniques typically protect only a single logic cone or a very small part of the circuit, leaving most of the logic unprotected. A designer is thus forced to put additional effort to identify the most impactful logic that, upon being protected, leads to maximal impact on the circuit output.
2. **High implementation overhead**. Attempts to maximize the output corruptibility or maximize the percentage of outputs protected by increasing the number of key gates or deploying multiple protection units can exacerbate the area, power, and timing overhead. Moreover, use of "expensive" one-way functions, as in case of ORF-Lock, inherently leads to an impractical overhead for logic locking.
3. **Reliance on logic synthesis**. SFLL does not offer any security guarantees regarding the functionality-strip operation. The same applies to ORF-Lock where secure merging of the one-way function with the original circuit is required. Thus, there is a strong need to develop secure logic synthesis algorithms that can modify/strip-away the desired functionality while ensuring that no traces are left in the circuit for the attackers to exploit.
4. **Attacks without oracle access**. While many logic locking attacks assume access to the oracle (functional IC), the de-synthesis attack and the SPS attack demonstrate that the significant information may be extracted from the locked netlist even without the access to a functional IC. More powerful oracle-less attacks may be launched in the future.
5. **Assumptions about the scan chains**. Bulk of the logic locking research has focuses on combinational circuits. The basic assumption is that with scan access, a sequential circuit may be treated as a combinational one; any combinational logic locking defense that is secure against attacks with full scan access is more powerful than and preferable over sequential logic locking defenses that assume a weaker threat model. Only a handful of research efforts focus on sequential

locking, where a circuit is locked by introducing additional states into the finite state machine [4]; the registers in the FSM are not part of the scan chains. In many realistic scenarios, only a fraction of flip-flops may be included in the scan chain; such scenarios have rarely been examined from a security stand-point [5].

11.3   Directions for Future Research

To tackle the aforementioned challenges, we anticipate the following research directions for logic locking:

1. **Leveraging cryptographic primitives**. Existing logic locking approaches deliver security against specific attacks, however, there is still the need for a system that is end-to-end secure. Significant research has been carried out in the field of program obfuscation, with the seminal results proving that program obfuscation is impossible in general, whereas certain classes of Boolean functions such as the point functions can be obfuscated securely [20]. Logic locking techniques can build upon the algorithms, metrics, and primitives utilized in the context of program obfuscation to establish security guarantees.
2. **Developing secure logic synthesis algorithms**. As already mentioned, there is a strong need to develop secure logic synthesis algorithms that offer provable security guarantees. Metrics to quantify the resilience of these synthesis algorithms are also required. Recent research has explored the use of ROBDDs for secure logic synthesis [6]; however, the overhead associated with ROBDDs can be prohibitive. Further research on secure and cost-effective logic synthesis is highly warranted. An interesting research direction to explore is the use of high level synthesis techniques in the context of security [8].
3. **Combining combinational and sequential logic locking.** Research on sequential [2, 4] and combinational logic locking has, more or less, been pursued as two independent research streams, perhaps due to the adoption of different analysis tools, techniques, and metrics. One possible research direction to synergistically combine the two types of logic locking. Integrated logic locking techniques may take into the metrics for both kinds of locking, thwarting attacks of both combinational and sequential nature.
4. **Integration with other DfTr techniques.** In this book, we have explored logic locking as a standalone solution for protecting against piracy from both untrusted foundries and end-users with reasonable implementation cost. Integration of logic locking techniques with other DfTr techniques such as camouflaging and split manufacturing might be beneficial in terms of overall security and implementation cost.
5. **Secure scan chain obfuscation.** Many attacks such as the SAT and sensitization attack rely on scan access to treat sequential circuits as combinational. A potential countermeasure to these attacks is to lock/obfuscate the scan chain(s) [3, 19]. However, such countermeasures have already been broken [1].

None of these techniques offer provable security guarantees. Thus, there is a need to revisit security of such obfuscation techniques and develop formally secure countermeasures.

6. **Securing analog and mixed-signal circuits**. Until recently, logic research had focused only on protecting digital circuits. However, analog and mixed-signal circuits are also an essential component of all electronic systems, and there exists a dire need for research on protecting these circuits. The earliest research effort in this emerging research area focus on locking analog current mirrors using an array of key controlled transistor arrays [18].

7. **Novel Applications of logic locking**. Until recently, logic locking has mainly been used to protect against piracy. However, the same technique can be applied to achieve other objectives. One such application is "performance locking", where instead of functionality, the performance/throughput of a chip is locked. Even with an incorrect key, the chip remains functional albeit at a lower performance [30]. Only upon application of the correct key, the chip offers the maximum throughput. Similarly, the concepts of logic locking could be extended to enable/disable selected modules in a chip, effectively supporting digital rights management.

This book focused on logic locking as an effective countermeasure against the emerging threats of IP piracy, overbuilding, and reverse engineering, which pose a risk of billions of dollars to the semiconductor industry. By locking a design during the untrusted manufacturing, test, and assembly stages, logic locking offers versatile protection against many attacks. Apart from receiving significant attention from academia over the past decade, logic locking is gaining footholds in the industry. For example, Mentor Graphics TrustChain framework supports basic logic locking [7, 16]. We can anticipate support for countermeasures against SAT and other attacks in the upcoming versions of such security platforms. However, the field of logic locking may still be considered in infancy, and a number of research questions remain open for further exploration.

References

1. Alrahis L, Yasin M, Saleh H, Mohammad B, Al-Qutayri M, Sinanoglu O (2019) ScanSAT: unlocking obfuscated scan chains. Cryptology ePrint Archive, Report 2019/005. https://eprint.iacr.org/2019/005
2. Baumgarten A, Tyagi A, Zambreno J (2010) Preventing IC piracy using reconfigurable logic barriers. IEEE Des Test Comput 27(1):66–75
3. Karmakar R, Chatopadhyay S, Kapur R (2018) Encrypt flip-flop: a novel logic encryption technique for sequential circuits. arXiv preprint arXiv:180104961
4. Koushanfar F (2012) Provably secure active IC metering techniques for piracy avoidance and digital rights management. IEEE Trans Inf Forensics Secur 7(1):51–63
5. Massad M, Garg S, Tripunitara M (2015) Integrated circuit (IC) decamouflaging: reverse engineering camouflaged ICs within minutes. In: Network and distributed system security symposium

6. Massad M, Zhang J, Garg S, Tripunitara M (2017) Logic locking for secure outsourced chip fabrication: a new attack and provably secure defense mechanism. CoRR abs/1703.10187. http://arxiv.org/abs/1703.10187
7. Mentor (2018) TrustChain^{TM} security platform. https://www.mentor.com/products/sm/trustchain-security-platform
8. Pilato C, Regazzoni F, Karri R, Garg S (2018) TAO: techniques for algorithm-level obfuscation during high-level synthesis. In: IEEE/ACM design automation conference, p 155
9. Plaza S, Markov I (2015) Solving the third-shift problem in IC piracy with test-aware logic locking. IEEE Trans CAD Integr Circuits Syst 34(6):961–971
10. Rajendran J, Pino Y, Sinanoglu O, Karri R (2012) Security analysis of logic obfuscation. In: IEEE/ACM design automation conference, pp 83–89
11. Rajendran J, Zhang H, Zhang C, Rose G, Pino Y, Sinanoglu O, Karri R (2015) Fault analysis-based logic encryption. IEEE Trans Comput 64(2):410–424
12. Roy J, Koushanfar F, Markov IL (2010) Ending piracy of integrated circuits. IEEE Comput 43(10):30–38
13. Shamsi K, Li M, Meade T, Zhao Z, Pan DZ, Jin Y (2017) Cyclic obfuscation for creating sat-unresolvable circuits. In: ACM great lakes symposium on VLSI, pp 173–178
14. Shamsi K, Li M, Meade T, Zhao Z, Z D, Jin Y (2017) AppSAT: approximately deobfuscating integrated circuits. In: IEEE international symposium on hardware oriented security and trust, pp 95–100
15. Shen Y, Zhou H (2017) Double DIP: re-evaluating security of logic encryption algorithms. Cryptology ePrint Archive, Report 2017/290. http://eprint.iacr.org/2017/290
16. Skudlarek JP, Katsioulas T, Chen M (2016) A platform solution for secure supply-chain and chip life-cycle management. Computer 49(8):28–34
17. Subramanyan P, Ray S, Malik S (2015) Evaluating the security of logic encryption algorithms. In: IEEE international symposium on hardware oriented security and trust, pp 137–143
18. Wang J, Shi C, Sanabria-Borbon A, Sánchez-Sinencio E, Hu J (2017) Thwarting analog IC piracy via combinational locking. In: IEEE international test conference, pp 1–10
19. Wang X, Zhang D, He M, Su D, Tehranipoor M (2017) Secure scan and test using obfuscation throughout supply chain. IEEE Trans Comput Aided Des Integr Circuits Syst 37(9):1867
20. Wee H (2005) On obfuscating point functions. In: ACM symposium on theory of computing, pp 523–532
21. Xie Y, Srivastava A (2016) Mitigating SAT attack on logic locking. In: International conference on cryptographic hardware and embedded systems, pp 127–146
22. Xu X, Shakya B, Tehranipoor M, Forte D (2017) Novel bypass attack and BDD-based tradeoff analysis against all known logic locking attacks. In: International conference on cryptographic hardware and embedded systems, pp 189–210
23. Yasin M, Mazumdar B, Ali SS, Sinanoglu O (2015) Security analysis of logic encryption against the most effective side-channel attack: DPA. In: IEEE international symposium on defect and fault tolerance in VLSI and nanotechnology systems, pp 97–102
24. Yasin M, Mazumdar B, Rajendran J, Sinanoglu O (2016) SARLock: SAT attack resistant logic locking. In: IEEE international symposium on hardware oriented security and trust, pp 236–241
25. Yasin M, Mazumdar B, Sinanoglu O, Rajendran J (2016) Security analysis of anti-SAT. IEEE Asia and South Pacific design automation conference, pp 342–347
26. Yasin M, Rajendran J, Sinanoglu O, Karri R (2016) On improving the security of logic locking. IEEE Trans CAD Integr Circuits Syst 35(9):1411–1424
27. Yasin M, Saeed SM, Rajendran J, Sinanoglu O (2016) Activation of logic encrypted chips: pre-test or post-test? In: Design, automation test in Europe, pp 139–144
28. Yasin M, Mazumdar B, Sinanoglu O, Rajendran J (2017) Removal attacks on logic locking and camouflaging techniques. IEEE Trans Emerg Top Comput 99(0):PP
29. Yasin M, Sengupta A, Nabeel MT, Ashraf M, Rajendran J, Sinanoglu O (2017) Provably-secure logic locking: from theory to practice. In: ACM/SIGSAC conference on computer & communications security, pp 1601–1618

30. Zaman M, Sengupta A, Liu D, Sinanoglu O, Makris Y, Rajendran JJV (2018) Towards provably-secure performance locking. In: IEEE design, automation & test in Europe conference & exhibition, pp 1592–1597
31. Zhou H, Jiang R, Kong S (2017) CycSAT: SAT-based attack on cyclic logic encryptions. In: IEEE/ACM international conference on computer-aided design, pp 49–56

Appendix A
Background on VLSI Test

A.1 Manufacturing Test

Each manufactured IC passes through a manufacturing test, which screens out defective chips. Defects, such as shorts or opens, may arise due to imperfections in the IC fabrication process. As illustrated in Fig. A.1, the test is conducted using automated test equipment (ATE), which applies test patterns to an IC and captures the output responses. An IC is marked defective if the observed responses differ from the expected ones. The test may be functional or structural. Functional testing generates test patterns based on the design specifications alone. Structural testing, however, takes into account the implementation level details of the circuit, achieving better test quality and lower test pattern count.

A.2 Fault Models

Fault models are mathematical representations of physical defects. A fault captures the impact of a defect on the logic function of an IC. Various models including single stuck-at, bridging, and delay fault model have been developed to encompass different classes of defects. The single stuck-at fault model remains the most prevalent fault model; the same model has been used throughout this book. The model assumes that the fault location is tied to a logical value (0 or 1), and at most a single fault can be present in the IC. Since each fault location can be either stuck-at-0 (s-a-0) or stuck-at-1 (s-a-0), the test patterns (which are generated by an ATPG algorithm) target both types of faults at each fault location.

© Springer Nature Switzerland AG 2020
M. Yasin et al., *Trustworthy Hardware Design: Combinational Logic Locking Techniques*, Analog Circuits and Signal Processing,
https://doi.org/10.1007/978-3-030-15334-2

Fig. A.1 Testing a circuit
using an ATE

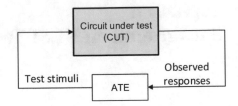

A.3   Automatic Test Pattern Generation (ATPG)

The objective of an ATPG algorithm is to generate test patterns for a given circuit. In order to minimize the test application time and maximize the test quality, ATPG algorithms aim at "detecting" (explained in Sect. A.4) the maximum number of faults with the minimal number of test patterns. While the traditional ATPG algorithms, such as D-algorithm, PODEM, and FAN, focus on the structural properties of a circuit, modern ATPG algorithms tend to make use of techniques such as Boolean satisfiability. The ATPG algorithms can be applied directly to and scale well for combinational circuits. However, the computational complexity of these algorithms increases significantly for sequential circuits, where it is difficult to control and/or observe the internal signals because of the presence of memory elements. For efficiently testing sequential circuits, structures such as scan chains are used that allow a sequential circuit to be treated as combinational, as explained in Sect. A.5.

A.4   Detection of a Stuck-at Fault

A stuck-at fault f may manifest as an error in the circuit output for certain input patterns. When an input pattern renders the impact of the fault f to be observable through the circuit output, the fault is said to detectable. Detection of a stuck-at fault involves two steps: *fault activation* and *fault propagation*. Fault activation necessitates setting the fault location to a value opposite to that of the stuck-at value, e.g., a value of 1 to detect a *s-a*-0 fault. Fault propagation entails forwarding the effect of the fault along a sensitization path to a primary output. Note that sensitization of a net to an output represents a bijective mapping between the two, i.e., the value of the net becomes observable on the output, as is or in the negated form. An input-output pair that detects a given fault by accomplishing both fault activation and propagation is referred to as a *test pattern*.

Example Consider the netlist in Fig. A.2. The s-a-0 fault on net h—denoted as h/0— is activated by setting the net h to 1. Note that faults may be represented as net/stuck-at value; e.g., c/0 represents a stuck-at-0 fault on net c. To propagate the fault to the primary output O2, the net i must be set to 0. An input pattern that detects the fault

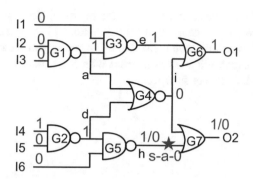

Fig. A.2 Detecting a s-a-0 fault on the net h

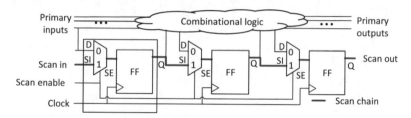

Fig. A.3 An example scan chain with three scan cells. The flip-flops are configured into scan cells by adding multiplexers at the flip-flop inputs. The Scan enable (SE) signal selects between the shift operation and normal operation

is 000100. The output O2 will be 1 in the fault-free circuit and 0 in the presence of h/0 fault; this is represented using the notation 1/0.

A.5   Scan-Based Testing

Design-for-testability (DfT) structures are inserted in an IC, early in the design cycle, to enable high-quality testing. The most commonly deployed DfT structures are scan chains. In scan-based testing, the flip-flops in a design are reconfigured as scan cells. Effectively, every flip-flop becomes controllable and observable through shift operations. Consequently, test generation algorithms can treat the flip-flops as inputs and outputs; the sequentiality is therefore eliminated, enabling the use of combinational test generation algorithms at reduced computational complexity. Note that the SAT, sensitization, and other attacks discussed in this book assume scan access.

Example As shown in Fig. A.3, each scan cell comprises a flip-flop and a multiplexer. The select line of the multiplexer (scan enable (SE) signal) decides the inputs to the flip-flops. When SE = 1, the flip-flops behave as a shift register; each flip-flop is loaded with the output value of the previous flip-flop in the scan chain. Test operations involve (1) scanning in the test pattern, (2) capturing the response of the combinational logic into scan cells, and (3) scanning out the captured response. The scanned-out response is then compared with the expected response to decide whether the IC is defective or functional.

Printed in the United States
By Bookmasters